SOTHIS PRESS

Sothis Press aims to completely reimagine the public domain book with an emphasis on universal access to captivatingly designed print editions. We believe that technology will enable every single book ever published to be perpetually accessible in beautiful print form. Sothis is beauty in ubiquity.

www.sothis.press
1 Waterhouse Square, London, ECIN 2ST

The Air-Brake

A Practical Presentation of the Modern Developments of the Air-Brake for Steam and Electric Railroad Trains

Llewellyn V. Ludy

SOTHIS PRESS

ISBN 978-1-0368-0983-6

The Air-Brake is in the public domain.

This edition is published by Sothis Press © 2023. The company is registered in England and Wales. The Company Number is 14900389.

THE AIR-BRAKE

A PRACTICAL PRESENTATION OF THE MODERN DEVELOPMENTS OF THE AIR-BRAKE FOR STEAM AND ELECTRIC RAILROAD TRAINS

By LLEWELLYN V. LUDY, M. E.
PROFESSOR OF EXPERIMENTAL ENGINEERING, PURDUE UNIVERSITY.
AMERICAN SOCIETY OF MECHANICAL ENGINEERS

ILLUSTRATED

CHICAGO
AMERICAN SCHOOL OF CORRESPONDENCE
1913

CONTENTS

CONTENTS

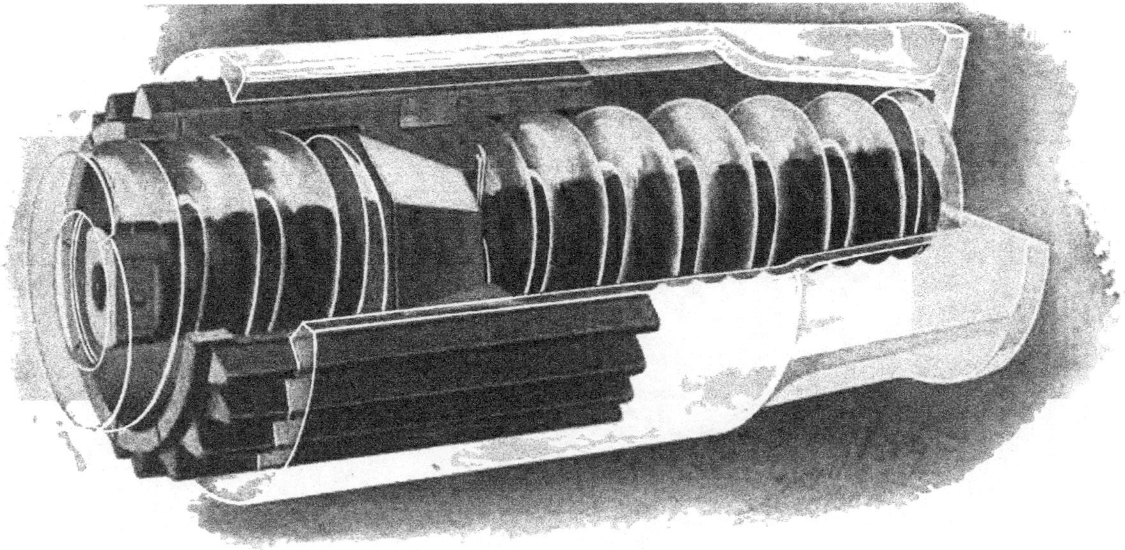

TRANSPARENT INTERIOR VIEW OF WESTINGHOUSE FRICTION DRAFT GEAR.

Application—The large "preliminary spring" at the left absorbs ordinary stresses and forces the friction strips against the friction cylinder. The small spring gives additional pressure on the wedge. When both springs become compressed, the friction strips are moved with increasing resistance, absorbing a pressure of over 150,000 lbs. before completion of stroke.

Release—The small spring at the left releases the wedge and the large spring returns friction strips to normal position, practically without recoil.

Representative American Locomotive, Built by the American Locomotive Company.

INTRODUCTION

WHEN George Westinghouse in 1869 invented the air-brake and thereby enormously increased the possibilities of control of railroad trains, he removed one of the most effective bars to high speed and freedom from accident for rapidly moving carriers. Lack of positive and reliable control leads always to destructive results, and therefore the railroads have been untiring in their efforts to stimulate the perfection of the air-brake as one of their most important appliances. Many roads at once adopted the device although it was still far from perfect. Three years later, the inventor added the triple valve, a distinct advance in reliability and facility of operation, and all of the systems which have survived are based on a variation of this method of action. Within the comparatively short period of twenty-five years, the application of the air-brake even to freight cars has become well-nigh universal, resulting in a much more expeditious handling of freight and in a greater safety to life and property.

¶ The present work is designed to emphasize the practical elements of design and construction, giving particularly the details of the Westinghouse and New York systems as applied to railroad equipment; a short presentation is also given of the air-brake as applied to electric cars. As the material was written especially for the correspondence courses of the American School, it should appeal both to the trained man who desires accurate information, and to those readers who are only interested in keeping up with the world's progress along technical lines.

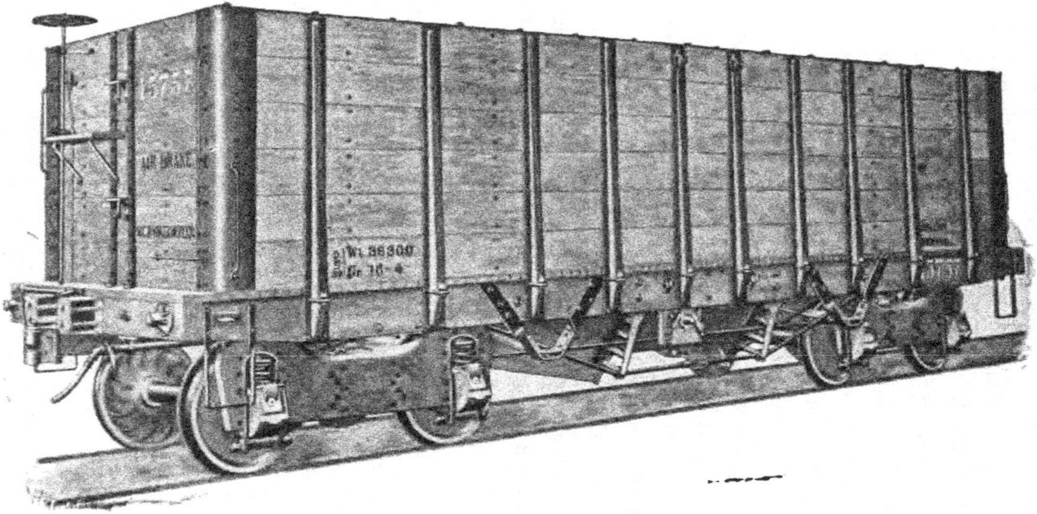

MODERN AMERICAN HOPPER-BOTTOM GONDOLA CAR, WITH AIR BRAKE EQUIPMENT.

THE AIR-BRAKE

INTRODUCTION

The development of the many accessory appliances with which the rolling stock of our railways is fitted, has been the subject of a great deal of study and investigation. Of the many appliances which have received careful and systematic study, the braking apparatus may be mentioned as one of the most important.

The time when the question of braking first received attention dates back further than the time when highways became sufficiently well made and well maintained to permit of vehicles being drawn at a moderate rate of speed. When wheeled vehicles, drawn at speeds of 10 or 15 miles per hour, first made their appearance, it was found necessary to provide means by which they could be easily and quickly stopped in case of an emergency. The first carts and wagons, built for agricultural purposes, were of such construction that the resistance of the earth and axle were sufficient to bring them to rest in a reasonable length of time on ordinary roads. In cases of steep grades, the motion was retarded by one or both wheels being locked with chains, or by a stone or piece of timber being chained to the axle and dragged along the ground behind the vehicle.

It is interesting to note that the question of braking has steadily increased in importance as the demand for higher speed has increased. This applies equally well to all classes of vehicles, including railway trains, street and interurban cars, automobiles, and wagons. The first forms of braking apparatus adopted have formed the basis of almost all brake appliances which have since been employed for the same class of vehicles.

Early Forms of Brake. Perhaps one of the first forms of brake used was that found on the early stage-coach. It consisted of an iron shoe which was chained to the fore part of the coach, and was used only on steep grades. To apply this brake, the shoe was removed from its hook under the carriage and placed on the ground in front of the rear wheel, in such a position that the wheel would roll on it. As the wheel rolled on the shoe, the chain became taut, with the result that both the shoe and wheel slid over the surface of the ground.

A railroad is known to have existed as early as 1630, although it would hardly be called by that name to-day. The construction of the track as well as that of the cars, was almost entirely of wood. Even with this crude construction, it was found necessary to provide a brake to control the speed of the cars on the slight grades. The form of brake devised to meet the conditions consisted of a wooden lever pinned to the frame of the car at one end, in such a manner as to permit of its being pressed against the tread of the wheel by hand. When not in use, the lever was held off the wheel by means of a chain. The principle employed here in resisting the motion of the car, is the same as that employed to-day on all railroads—namely, of *applying the braking or resisting force to the periphery of the wheel.*

As railroads increased in number and their construction improved, braking apparatus became more and more a necessity. As a result, inventors brought out a number of simple braking appliances. The question of braking, however, did not become a very important or serious one until the advent of the steam locomotive. Previous to its coming, the cars were small and were drawn by animals, and the speeds were low; but with the steam locomotive in existence, an efficient brake became an absolute necessity.

This problem received the close attention of inventors and investigators; and at the close of 1870, the *automatic, electromagnetic, steam, vacuum,* and *air brakes* were found in use on the railroads in the United States. These types of brakes differed chiefly in the manner in which the braking power was obtained. Other devices were invented, but could not stand the test of actual practice and did not come into prominence.

It might be interesting to note briefly one or two rather unique types of brakes not included in any class yet mentioned. The *Cramer Brake,* brought out in 1853, might be mentioned as one of these. Its principal feature consisted of a spiral spring which was connected to the brake-staff at the end of each car. This spring was wound up by the brakeman before leaving the station. The brake apparatus on each car was under the control of the engineer, through the medium of a cord. This cord was connected to the mechanism of each brake, and passed through the cars, terminating in the cab on the engine. The engineer, desiring to stop his train, would shut off the steam and give the cord a pull, which action resulted in releasing the coil springs on the various cars, and applied the brakes by winding up the brake-chains.

The *Loughridge Chain Brake* is another unique brake, which was introduced in 1855. The Loughridge brake consisted of a combination of rods and chains which extended from a winding drum under the engine, throughout the entire length of the train. This continuous chain joined other chains under each car, which in turn were connected to the brake-levers. The winding drum located under the engine was connected by a worm gear to a small friction wheel. In operating the brake, a lever in the cab was thrown, which brought the small friction wheel in contact with the periphery of one of the driving wheels, thereby causing the drum to rotate and wind up the chain. The movement of the chain, which was experienced throughout the entire length of the train, served to actuate levers and rods under each car, which in turn applied the brake-shoes to the treads of the wheels.

The early types of hand-brakes underwent many improvements as years went on and as experience demanded. Although during many years of early railroading, the braking on all trains was done by hand, nevertheless there was a constant desire and demand for a practical automatic brake. The rather crude and inefficient types of brakes already referred to were obtained only after a great many failures. Since about 1870, all forms of brakes have differed chiefly in but one respect—namely, in the appliances which are used in operating the foundation brake-gear. The foundation brake-gear is made up of the rods, levers, pins, and beams, located under the frame of the car, the operation of which causes the brake-shoes to be pressed against the periphery or tread of the wheel. The present scheme of applying the brake-shoe to the periphery of the car wheel—which was in use long before the first locomotive made its appearance—later experience has proven to be the most practical.

Many forms of brakes were devised prior to the year 1840; but, at that time, few locomotives were equipped with braking apparatus. About this period, however, when the locomotive tender began to take on some definite form, we find the tender fitted with braking appliances. Previously, when brakes were provided, they were usually found fitted to the cars only. It is only within the last thirty-five years that locomotives have been built with brakes fitted to the drivers. To-day it is not uncommon to find all wheels on both the locomotive and the tender equipped with braking apparatus.

In 1869, the first Westinghouse air-brake made its appearance. This brake is now referred to as the *Straight Air-Brake.* It was not an automatic brake. It consisted chiefly of a steam-driven air-compressor and storage reservoir located on the engine; a pipe line extending from this reservoir throughout the length of the train; a brake-cylinder on each car; and a valve located in the cab for controlling the brake mechanism. The train line was connected between cars by means of flexible rubber hose with suitable couplings. Each car was fitted with a simple cast-iron brake-cylinder and piston, located underneath the frame, the piston-rod of which connected with the brake-rigging in such a manner that when air was admitted into the cylinder, the piston was pushed outward and the brake thereby applied. In operating the brake, air was admitted into the train line from the storage reservoir by means of a three-way cock located in the cab. The air was conducted to the brake-cylinder under the various cars by means of the train-pipe. The release of the brakes was accomplished by discharging the air in the various brake-cylinders and the train-pipe, into the atmosphere, through the three-way cock in the cab. This was the simplest and most efficient brake invented up to the time of its appearance, and was adopted by many railroad companies in this country.

The *Straight Air-Brake* system, however, possessed three very objectionable features: *First,* in case of a break-in-two, or of a hose bursting, the brake at once became inoperative; *second,* it was very slow to respond in applying and releasing the brakes; and, *third,* the brakes on cars nearest the engine were applied first, causing jamming and surging of the cars, which sometimes proved destructive to the equipment. In order to overcome these undesirable qualities, Mr. George Westinghouse invented the *Westinghouse Automatic Air-Brake* in 1872. This form of brake, which has since gone out of service on steam railroads, was known as the *Plain Automatic Air-Brake.* This brake retained the principal features of the *Straight Air-Brake;* but, in addition, each car was provided with an air-reservoir, which supplied air for operating its particular brake-cylinder. The charging with air of this *auxiliary reservoir*, the admitting of this air into the brake-cylinder, and the discharge of the air from the brake-cylinder to the atmosphere, were accomplished by an ingenious device known as the *triple valve*. A detailed description of this valve will be given later.

In this same year (1872), the *Vacuum Brake* was invented; but, on account of its many undesirable features, it never gained very great prominence in this country. This brake was spoken of as the *Plain Vacuum Brake*, and was followed later by the *Automatic Vacuum Brake*. The principal parts of the air-brake were, in general, embodied in the Vacuum brake. One marked difference existed, however, in that, instead of an air-compressor, an *ejector* was installed on the locomotive, which exhausted the air from the train-pipe when the system was in operation.

At the close of the year 1885, there could be found in use on the railroads of the United States a number of different types of brakes. These could be grouped into two general classes—*Continuous* or *Air* brakes, and *Independent* or *Buffer* brakes. In the Buffer brake, the brake-shoes were actuated by rods and levers, which in turn received their motion from the movement of the draw-bar. It is easily seen that, with such a variety of different forms of braking apparatus, it would be impossible to control a train properly if it were made up of cars from different railroads having different brake systems.

Interchangeable Brake System. The one agency which has had an important part in placing the braking appliances of our railroads on the present high standard of perfection, is the Master Car-Builders' Association. This Association, realizing the increasing demand for the interchange of cars, saw the need of interchangeable brake systems. It was principally through the research of their committees that the brake systems of to-day are interchangeable and efficient.

The first experiment conducted by the committee in 1886 clearly showed that any further attempt to use the Independent or Buffer brake was not desirable, on account of the severe shocks resulting when stopping the train. The effect of the report of the committee was the withdrawal of this type of brake from the attention of the railroad officials. This left almost the entire field open to the Continuous or Air brake system. The committee continued its work the following year, and, from the results of a large number of tests, reported that the best type of brake for long freight trains was one operated by air, in which the valves were actuated by electricity. This type of brake stopped the train in the shortest possible distance, reduced all attending shocks to a minimum, was released instantaneously, and could be applied gradually. Although the results of tests pointed to the superiority of the air-brake in which the valves were operated by electricity, yet to-day we find no such systems in general use.

From the time of these tests, the different brake companies turned their attention to the style of brake represented by the Westinghouse Automatic Air-Brake system. In this system, the most important parts are the *triple valve,* located on the brake-cylinder of each car, and the *controlling* or *engineer's brake-valve* located in the cab. By the year 1893, a number of triple valves and engineer's brake-valves had been placed on the market, and representative ones were exhibited at the Columbian Exposition in Chicago in that year.

The committee of the Master Car-Builders' Association, being conscious of the fact that the actions of the valves made by the different companies were so widely different, proposed a series of tests of triple valves, and asked the different companies to submit valves for the said tests. The object of the proposed tests was to obtain data from which could be formulated a code of tests for triple valves which would be satisfactory to all parties concerned. The ultimate aim of the committee was to secure triple valves which would operate with the same ultimate effect when subjected to identical conditions, and which would operate successfully when intermingled with each other in a train.

Such tests were conducted on a specially constructed air-brake testing track in the year 1894. Five companies responded with valves for the series of tests, of which the valves representing the Westinghouse and New York companies gave the best results. From the results obtained, the committee prepared a code of tests for triple valves, which code was soon after adopted by the Association as standard. As a result of this action, makers of air-brake apparatus endeavored to produce triple valves which would give results as specified in the code. This naturally led to interchangeable air-brake systems—one of the objects the committee hoped to attain. Many triple-valve tests have since been made, and the code has been changed from time to time to meet new conditions which have developed.

To-day there are mainly two air-brake systems in general use on steam railroads in this country, namely—the *Westinghouse* system and the *New York* system.

WESTINGHOUSE AIR-BRAKE SYSTEM

The principle on which the Westinghouse Air-Brake system operates, is, that if, after the system is charged with compressed air, a reduction is made in the brake-pipe pressure, the brake will be applied; and in order to release the brake, the brake-pipe pressure must be restored. It follows that if any accident occurs to the braking apparatus which reduces the pressure in the brake-pipe—such as the train parting or hose bursting—the brakes will at once be applied. In this respect, the Westinghouse Air-Brake system is automatic. The system is composed of the following principal parts:

1. The *steam-driven air-pump* located on the engine, which furnishes compressed air for the whole system.

2. The *main reservoir*, which is located some place about the engine or tender, and in which compressed air is stored at a pressure of 90 pounds.

3. The *engineer's brake-valve*, which is located in the cab, by means of which the flow of air from the main reservoir to the brake-pipe and from the brake-pipe to the atmosphere is regulated.

4. The *air-pressure gauge*, which is located in plain view of the engineer, and which contains two pointers, one red and one black. The black hand indicates the brake-pipe pressure; and the red hand, the main reservoir pressure.

5. The *pump-governor*, which regulates the flow of steam to the pump, shutting off the steam when the maximum pressure carried in the main reservoir is reached.

6. The *brake-pipe* which consists of a pipe under each car with flexible hose connection between cars, connecting all the triple valves with the engineer's brake-valve. This pipe has been, and is sometimes still, referred to as the *train line;* but as there are other train lines, such as the signal and steam lines, it has of late been referred to as the brake-pipe.

7. The *auxiliary reservoir* on each car, in which air is stored for use in applying the brake.

8. The *brake-cylinder* on each car, which contains a piston and rod. When air is admitted behind the piston, it causes it to move outward, and, by means of suitable connection to the foundation gear, applies the brake.

9. The *triple valve*, located under each car, the operation of which admits air from the brake-cylinder to the atmosphere, and recharges the auxiliary reservoir.

10. The *pressure-retaining valve*, which, when closed, will retain a pressure of 15 pounds in the brake-cylinder and thus prevent a complete release.

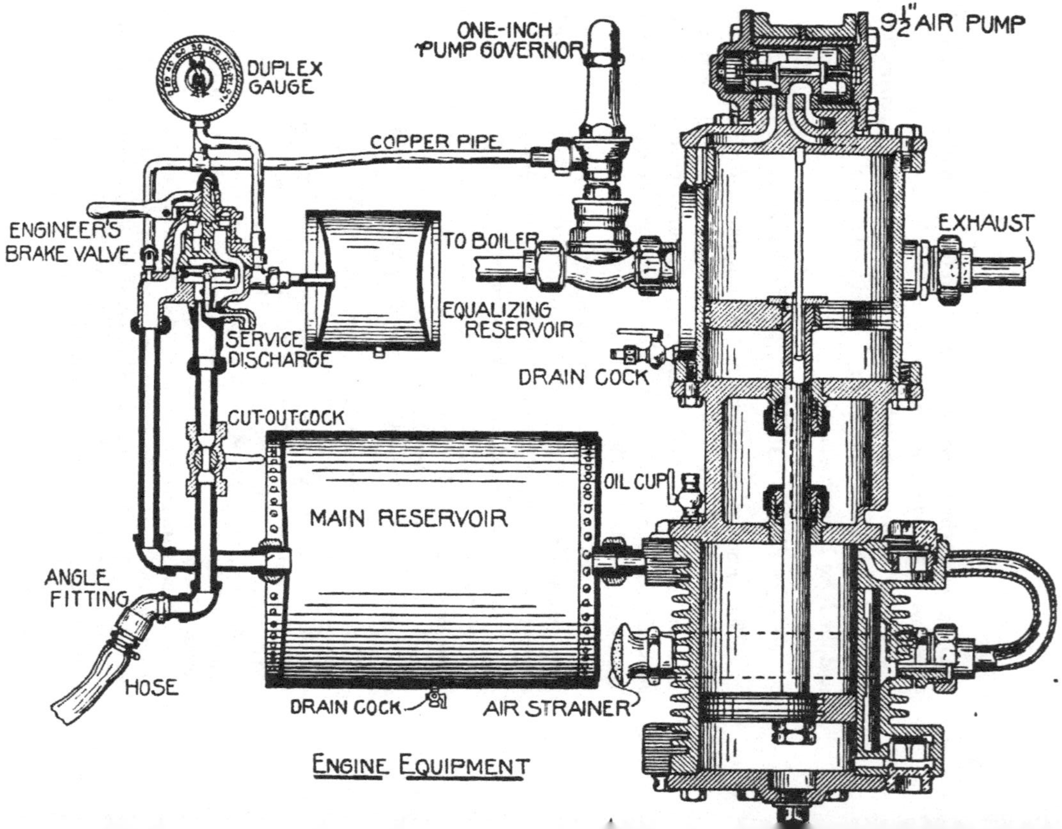

ENGINE EQUIPMENT

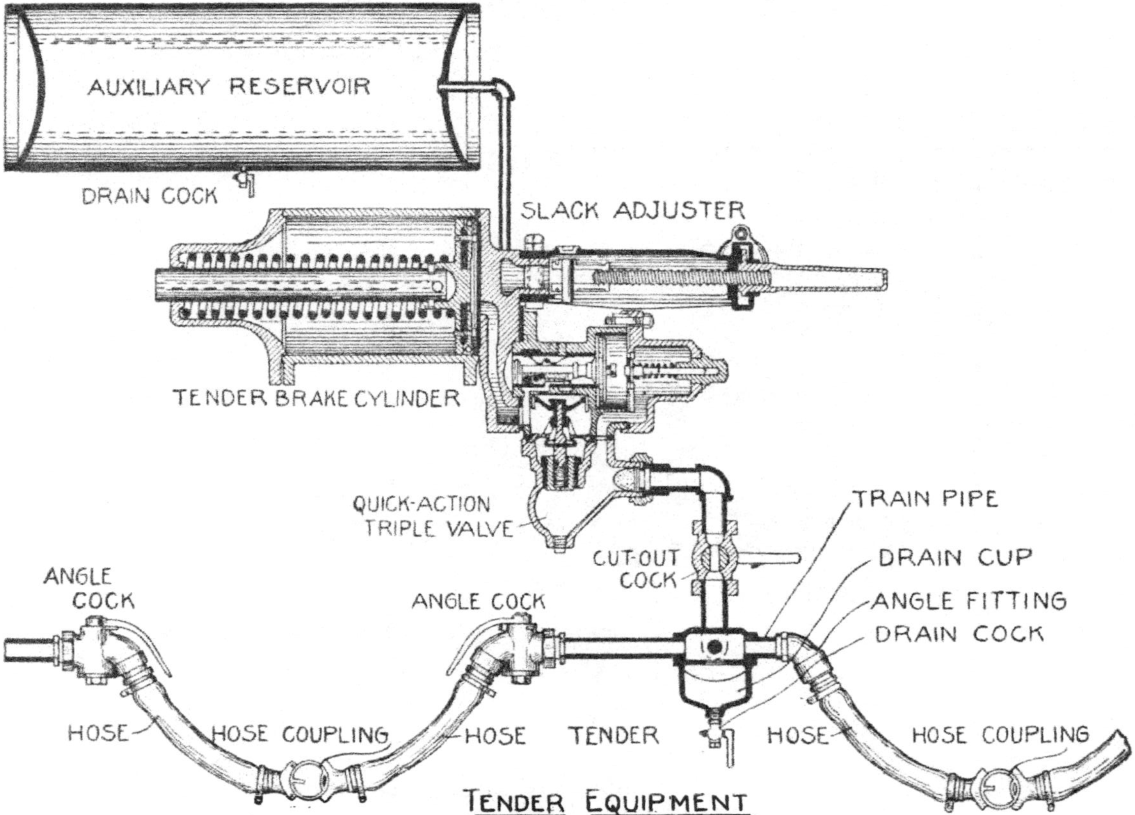

Fig. 15.—Tender Equipment, Westinghouse-Air-Brake System.

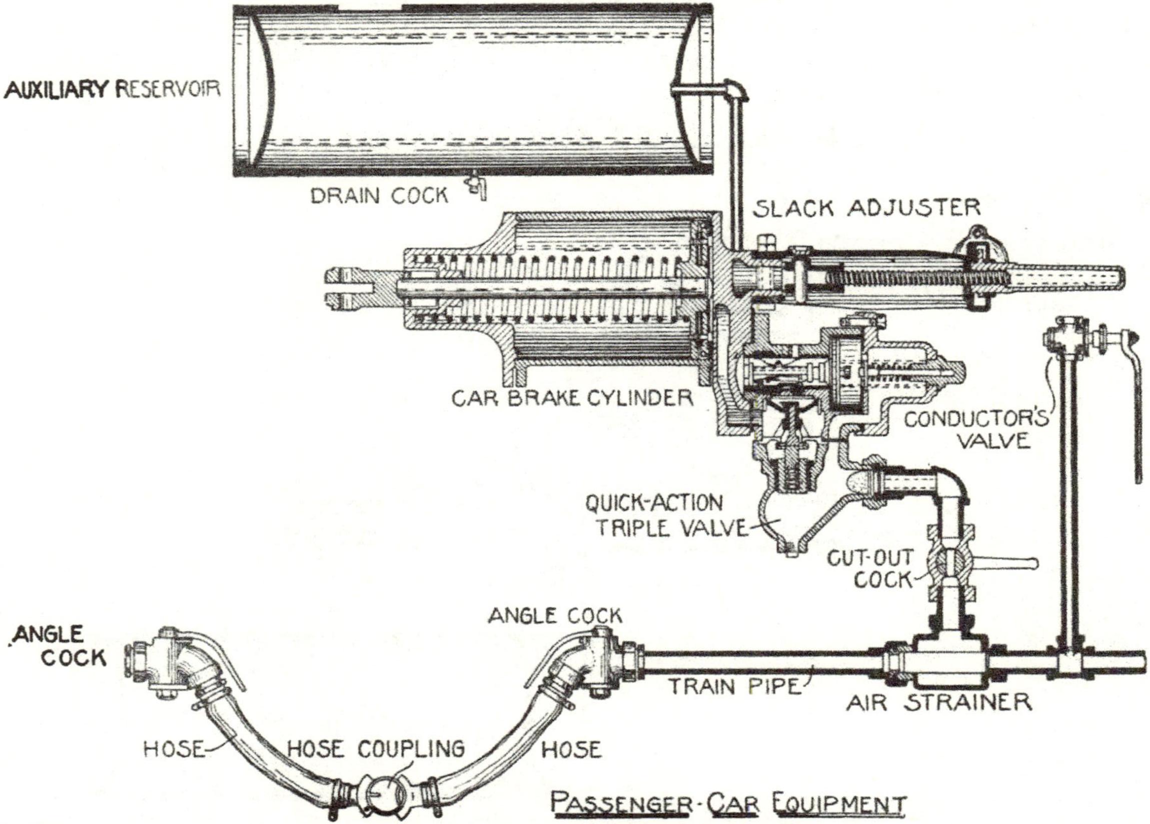

Fig. 1c. Passenger-Car Equipment, Westinghouse Air-Brake System.

The general arrangement of the Westinghouse air-brake system is shown in Fig. 1, (*a*, *b*, and *c*). This diagrammatically illustrates the arrangement and gives the names of the parts as found upon the locomotive, tender, and first car of a passenger train.

Operation of the Westinghouse Air-Brake. When the brakes are in operating condition, the pump-governor is set to maintain a pressure of 90 pounds in the main reservoir. This pressure is reduced by a valve attached to the engineer's brake-valve, which keeps the brake-pipe pressure at 70 pounds, when the engineer's brake-valve is in running position. The operation of the brake is controlled by the engineer's brake-valve, which has five fixed positions for its handle. These positions named in order, beginning from the left, are: *Release*, *running*, *lap*, *service*, and *emergency*.

When the engineer's brake-valve is in *running* position, a pressure of 70 pounds is maintained in the brake-pipe by means of the feed-valve. The brakes will release when the valve is in this position, but they will do so very slowly.

To make a *service* application of the brakes, the handle of the engineer's valve is placed in service position. In this position, the connection between the main reservoir and the brake-pipe, through the feed-valve, is closed. Air from the brake-pipe is allowed to escape to the atmosphere through ports in the valve. The brake-valve is left in this position only for a short time, when it is placed in lap position.

In the *lap* position, all working parts are closed and the brakes are held applied.

When it is desired to release the brake after either a service or an emergency application, the handle of the engineer's valve is placed in *release* position. In this position, direct connection is made between the main reservoir and the brake-pipe.

When it is necessary to make an *emergency* application, the handle of the engineer's brake-valve is placed in emergency position, and direct connection is made between the brake-pipe and the atmosphere. This causes a sudden reduction of pressure in the brake-pipe, and gives a higher pressure in the brake-cylinder than is obtained in service applications.

Westinghouse Nine and One-Half Inch Air-Pump. The nine and one-half-inch pump is shown in Figs. 2, 3, and 4. The pump consists of two cylinders. The lower (1) is the air-cylinder, and the upper (2) is the steam cylinder. The pistons of these cylinders are connected by the hollow rod (3). The air end is very simple, free air being drawn in at the valves (4), and discharged under pressure at the valves (5). The steam end is somewhat complicated in structure, but very simple in principle. The valve gear consists of two pistons (6) and (7), of unequal diameters, connected by the rod (8). This rod gives motion to a common D slide-valve (9) which admits steam above and below the main piston (10). The motion of the piston (6) and (7) is obtained by the slide-valve (11), Fig. 2, admitting and exhausting steam at the right of the piston (6). Steam enters the valve chamber *B* through the passage *A*. With the parts in the position shown in Figs. 2 and 4, steam from the chamber *B* passes down through the open port *C* below the piston (10), causing it to rise. The steam above the piston (10) passes out through the port *D* into the chamber *E* in the slide valve (9) through the exhaust port *F* to the atmosphere. The piston (10) continues to rise until the plate (12) strikes the upper shoulder on the rod (13) which raises the slide-valve (11). This allows the steam to pass from the chamber *B* through the port *H* around the slide valve (11) through the port *G* into the chamber *I*. In this position, the same pressure acts on both sides of the piston (6). Since the chamber *J* is always in communication with the exhaust passages through the port *K*, the pressure on the piston (7) is unbalanced. This causes all parts to move to the left. This movement opens the port *D*, permitting steam to force the piston (10) downward. At the same time, the port *C* is put in connection

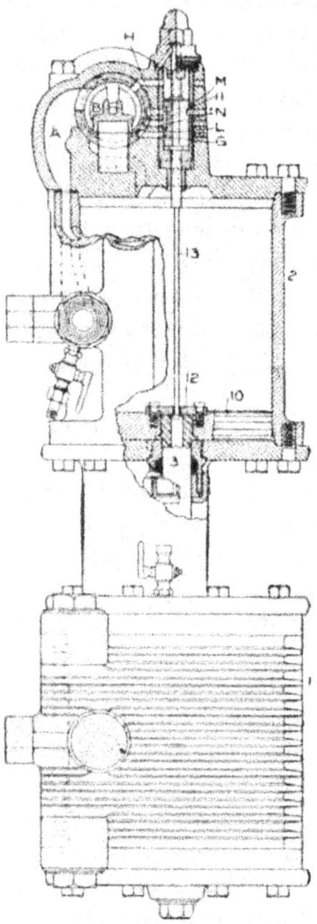

Fig. 2. Westinghouse 9½-Inch Air-Pump, End Section.

with the exhaust port *F* through the cavity *E* in the slide-valve (9). The piston (10) continues to fall until the plate (12) strikes the button on the lower end of the rod (13), drawing the valve (11) downward to the position shown in Fig. 2. This closes the port *G*, and connects the chamber *I* through the port *L*, the cavity *N*, and the port *M*, with the main exhaust *F*. This releases the pressure on the right of the piston (6). The pressure on the piston (6) now being greater than on the piston (7) will cause both to move to the right, thus completing the cycle.

Eight and One-Half Inch Cross-Compound. This pump is coming into use as a result of the growing demand for more air on long freight trains. Its capacity is about three and one-half times that of the 9½-inch pump above described. As illustrated in Fig. 5, this pump is of the duplex type, having two steam and two air cylinders arranged with the steam cylinders above and the air cylinders below. The high-pressure steam cylinder is 8½ inches in diameter; and the low-pressure, 14½ inches in diameter, both having a 12-inch stroke. The low-pressure air cylinder is 14½ inches in diameter, and is located under the high-pressure steam cylinder. The high-pressure air cylinder is under the low-pressure steam cylinder, and is 9 inches in diameter. The valve gear is located on the top head of the high-pressure steam cylinder, and is very similar to that of the 9½-inch pump already described. Figs. 6 and 7 show diagrammatically a cross-section through the pump. Fig. 6 is a diagram of the parts during an up-stroke of the high-pressure steam side, and Fig. 7 shows the down-stroke of the high-pressure steam side. The high-pressure steam piston is shown on the right, and the low-pressure on the left. The high-pressure steam piston, with its hollow rod, contains the reversing-valve rod, and operates the reversing valve in the same manner as that of the 9½-inch pump. This valve operates the main valve in the same manner as that described in the case of the 9½-inch pump. The main slide-valve controls the steam admission to, and the exhaust from, both the high- and low-pressure steam cylinders.

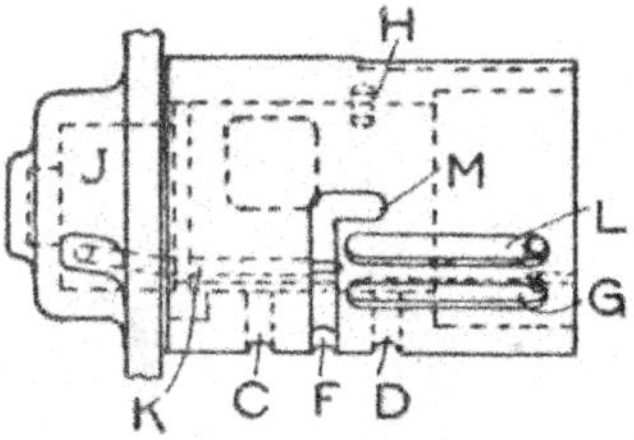

Fig. 3. Main Valve Bushing of Westinghouse 9½-Inch Air-Pump.

It is provided with an exhaust cavity, and in addition has four steam ports in its face. The two outer and one of the intermediate ports communicate with cored passages extending longitudinally in the valve, which serve to make the connection between the high- and low-

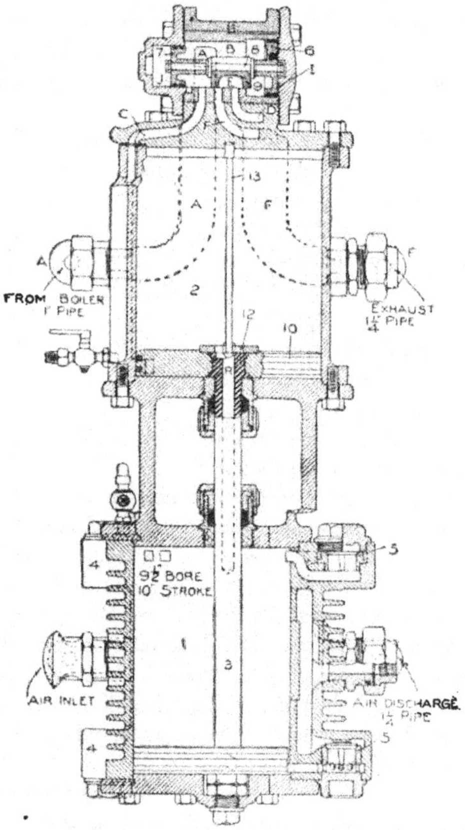

Fig. 4. Westinghouse 9½-Inch Air-Pump, Longitudinal Section.

pressure cylinders during the expansion of steam from one to the other. The other port controls the admission of steam to the high-pressure cylinder.

The valve seat has five ports. Of these, the two at the right, shown in Figs. 6 and 7, lead to the upper and lower ends of the high-pressure steam cylinder. The first and third from the left lead to the upper and lower ends of the low-pressure steam cylinder; and the second, to the exhaust. By following the arrows in Figs. 6 and 7, the flow of air and steam through the pump can easily be traced.

The principle of compounding employed in this pump enables it to compress air much more economically than is possible with the simple pump.

Main Reservoir. The use of the main reservoir is for storing an abundant air-supply to be used in charging and releasing the brakes. A large reservoir is of great importance, especially in freight service, since it provides air for an immediate recharging of the auxiliary reservoirs without running the pump intermittently at high rates of speed. The main reservoir should have a capacity of not less than 24,000 cubic inches on passenger engines, and not less than 40,000 cubic inches on freight engines. The main reservoir is usually located somewhere on the engine; but sometimes it is placed on the tender, though the latter location necessitates two extra pipe connections between the engine and the tender, which is not good practice. A good practice is to divide the main reservoir, and place half on each side under the running-board. The air is then delivered to one side and taken out of the other, the two reservoirs being connected. This system has two decided advantages over others, one being that the air is cooled, thus causing the moisture to be collected in the reservoir. The other advantage is, that the distance between the inflow and out-take prevents much of the dirt and oil from being

Fig. 5. Westinghouse 8½-Inch Cross-Compound Pump.

carried into the brake-pipe. The main reservoir should always be *drained* after each run.

Air-Pump Governor. The purpose of the governor is to cut off the steam supply to the pump when the desired main-reservoir pressure has been reached. Fig. 8 is a section through the governor. When the pump is in operation, steam enters the governor at *B*, passes the valve (2), and enters the pump through the pipe *C*. An air connection is made at *A* with the main reservoir. Main-reservoir air thus

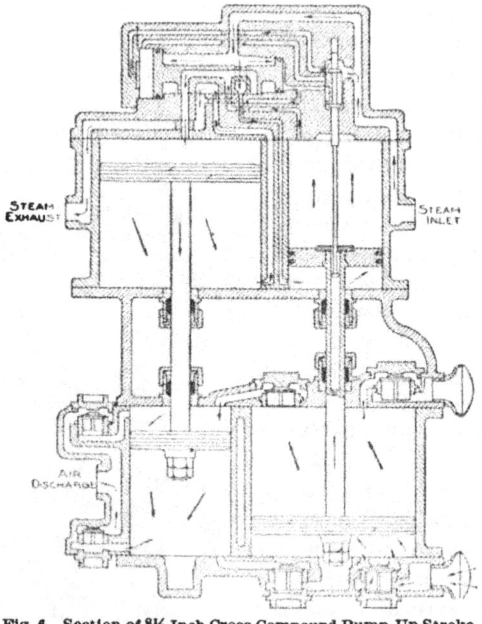

Fig. 6. Section of 8½-Inch Cross-Compound Pump, Up-Stroke, High-Pressure Steam Side.

enters the chamber *D;* and as soon as this pressure on diaphragm (1) is sufficient to overcome the tension of the spring (3), the diaphragm (1) will be raised and will unseat pin valve (4). Air will then flow down into the chamber *E,* forcing the piston (5) downward, thus seating the valve (2) and shutting off the steam from the pump. When the air-pressure falls below that carried in the main reservoir, the spring (3) will force the diaphragm (1) down, and will seat the pin valve (4). The air in the chamber *E* will now escape to the atmosphere through the small relief port *F*, and the spring (6) will open the valve (2), again admitting steam to the pump.

While the pin valve is unseated, there is a small escape of air to the atmosphere through the port *F*. This leakage, together with a leakage of steam through a small port in the valve (2), serves to keep the pump slowly operating and thus avoids trouble from condensation in the steam pipe.

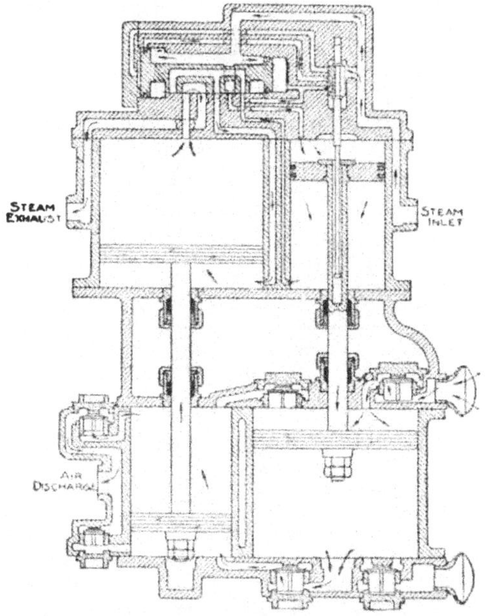

Fig. 7. Section of 8½-Inch Cross-Compound Pump, Down-Stroke, High-Pressure Steam Side.

Westinghouse Engineer's Brake-Valve. The construction of the engineer's brake-valve is illustrated in Figs. 9, 10, and 11. Fig. 9 is a section through the body, with the rotary valve removed, and shows the different positions of the handle. Figs. 10 and 11 are vertical sections of the entire valve taken at right angles to each other. The description of the operation of the valve is given in the order in which it is generally used when braking a train.

Running Position. The valve is shown in running position in Fig. 10. Main-reservoir air from the chamber *A* flows through the port *B* in the rotary valve (1) into the passage *C*, which conducts it to the feed-valve. (The course of the air through the feed-valve will be described later.) From the feed-valve, the air is conducted by the

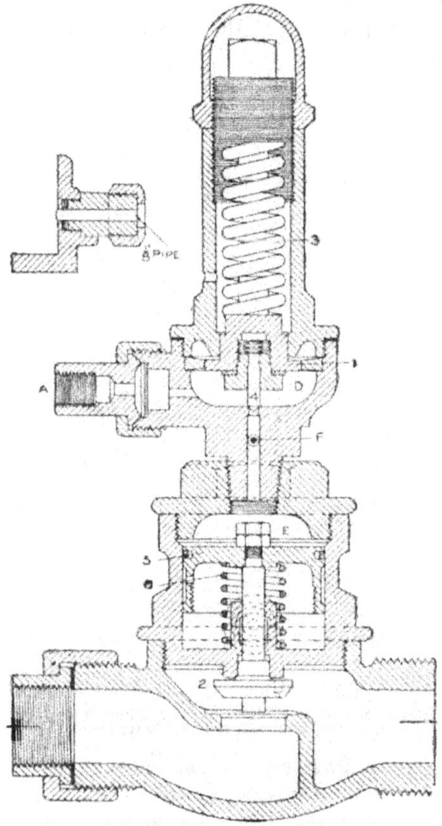

Fig. 8. Section through Air-Pump Governor.

passages *D* and *E* to the chamber *F*, and from here to the brake-pipe. The cavity *G* in the rotary valve (1) connects the passage *E* with the port *H*, permitting air at brake-pipe pressure to enter the chamber *I* and flow through the passage *J* into the equalizing reservoir.

Brake-pipe pressure now exists on both sides of the equalizing piston (2). Air continues to flow into the brake-pipe and equalizing reservoir until the pressure reaches 70 pounds. At this pressure, the feed-valve closes the passage leading from the main reservoir. In this position, the brake-pipe is kept charged to 70 pounds' pressure; 90 pounds' pressure is maintained in the main reservoir; and the entire system is ready for an application.

Service Position. When it is desired to reduce the speed of a train or to stop at a station, the handle of the engineer's brake-valve is placed in *service* position until the brake-pipe reduction causes enough air to enter the brake-cylinders to produce the desired result. When the brake-pipe reduction is sufficient, the handle of the valve is placed in *lap* position as described below. In service position

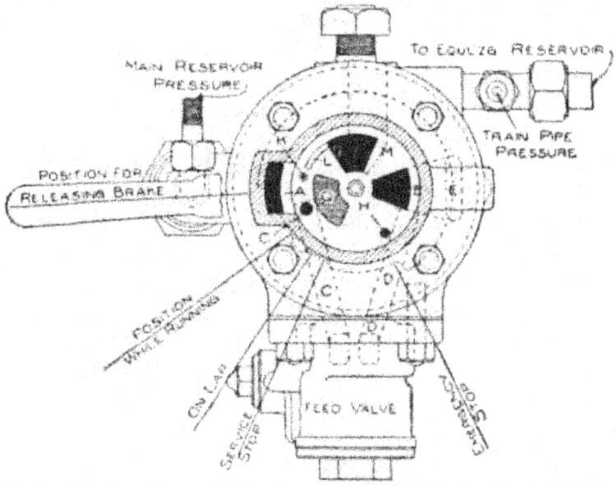

Fig. 9. Section through Engineer's Brake-Valve.—Rotary Valve Removed, and Different Positions of Handle Shown.

a groove in the rotary valve (1) connects the port *K* with the groove *L*, both of which are in the valve-seat. This permits air from the chamber *I* and the equalizing reservoir to discharge through the passage *M* to the atmosphere, thus reducing the pressure on the top side of the equalizing piston (2). Brake-pipe pressure, being greater than the pressure on the top side of the equalizing piston (2), forces it upward, opening the attached discharge valve, and permitting air to flow from the brake-pipe through the port *N* and the passage *O* to the atmosphere. When the pressure in the equalizing reservoir is reduced the desired amount, the handle of the engineer's brake-valve is moved to lap position. Air continues to discharge through the above-mentioned passages until the pressure in the brake-pipe has reduced slightly below that in the equalizing reservoir and chamber *I*; then the greater pressure acting on the top of the piston (2) causes it automatically to close the discharge valve. When the piston (2) closes the discharge valve, the pressure in the brake-pipe and equalizing reservoir is about the same.

The equalizing reservoir is 10 inches in diameter and 12 inches

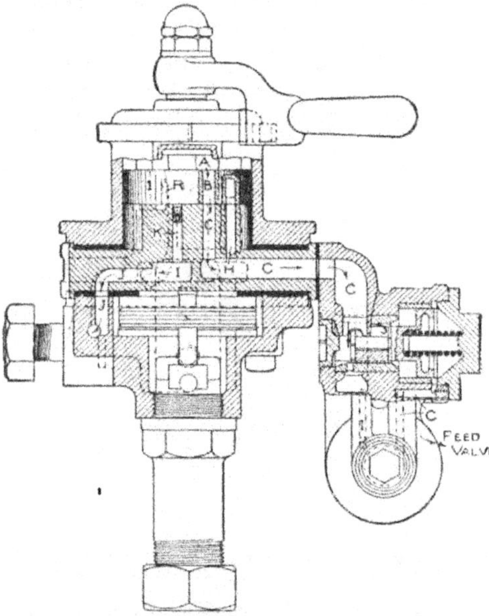

Fig. 10. Engineer's Brake-Valve in Running Position.

long. Its purpose is to increase the volume of air so that the pressure in the chamber *I* will not drop too rapidly when the handle is placed in service position. If a reduction of pressure be made in the equalizing reservoir, and the handle placed in lap position, air will exhaust from the brake-pipe until its pressure is the same as that in the equalizing reservoir.

Lap Position. This position is the one in which the valve handle is placed after a light reduction has been made in the brake-pipe.

It remains in this position, holding the brake applied until a further brake-pipe reduction is desired or until the brake is released. In this position, all ports are operatively closed. No air can enter the brake-pipe from the main reservoir, and no air from the brake-pipe can escape to the atmosphere.

Emergency Position. In case of an emergency, the handle of the brake-valve is moved to the extreme right. In this position, direct-application-and-exhaust port M is connected with direct-application-and-supply port E by the cavity G in the rotary valve (1). This establishes direct communication between the brake-pipe and the atmosphere, and a sudden reduction of pressure occurs in the

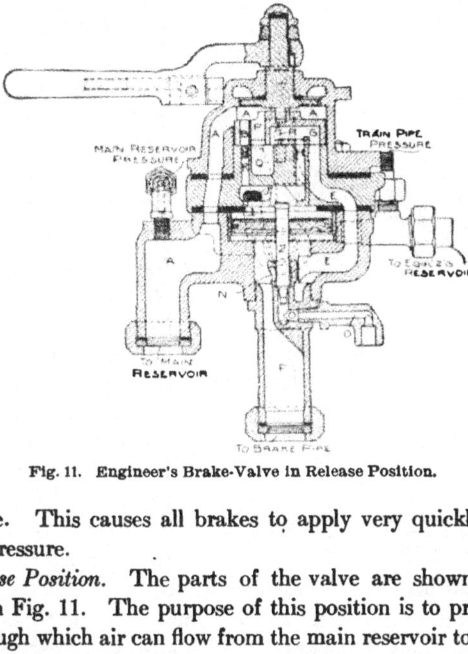

Fig. 11. Engineer's Brake-Valve in Release Position.

brake-pipe. This causes all brakes to apply very quickly with full braking pressure.

Release Position. The parts of the valve are shown in release position in Fig. 11. The purpose of this position is to provide large ports through which air can flow from the main reservoir to the brake-pipe and quickly recharge the system and release the brakes. Air from the main reservoir flows from the chamber A, through the port P, the cavities Q and G, into the passage E and to the brake-pipe. The ports H and K both conduct air to the chamber I, and the equalizing reservoir is charged through the passage *J* to the same pressure as exists in the brake-pipe.

The brake-valve handle should not remain in this position too long, as there is danger of overcharging the brake-pipe and the auxiliary reservoirs. The *warning port R* in the rotary valve (1) permits a small amount of air to escape from the chamber *A* to the exhaust passage *M*, making a noise which notifies the engineer that the brake-valve is still in release position.

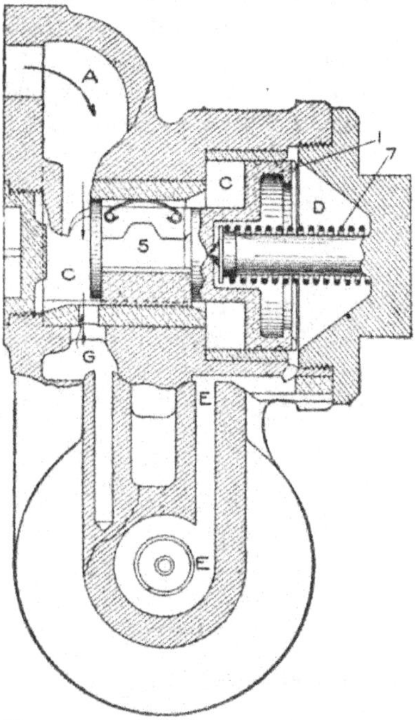

Fig. 12. Central Section through Supply-Valve Case and Governing Device of Slide-Valve Feed-Valve.

Slide-Valve Feed-Valve. The purpose of the feed-valve is to maintain a predetermined pressure in the brake-pipe while the engineer's valve is in running position. Figs. 12 and 13 illustrate the slide-valve feed-valve. Fig. 12 is a central section through the supply-valve case and governing device. Fig. 13 is a transverse section through the supply-valve case, and a central section through the regulating valve and spring-box. The ports *A* and *B* register with the ports in the brake-valve. The port *A* is in communication through the engineer's valve with the main reservoir, when the engineer's valve is in running position. Air enters the feed-valve at *A*, and has at all times free communication with the chamber *C*. The chamber *C* is separated from the chamber *D* by the supply piston (1). Connection is made between the chamber *D* and the brake-pipe through the passage *E*, the regulating valve (2), the chamber *F*, and the port *B*. The regulating valve (2) is normally held open by the tension of the spring (3) upon the diaphragm

(4). When the valve (2) is open, the chamber *D* has brake-pipe pressure acting upon it as described above. When the handle of the engineer's brake-valve is placed in running position, main-reservoir air enters the chamber *C* and forces the piston (1) to the right, as shown in Fig. 12. This draws the slide valve (5) to the right, uncovers the port *G*, and direct connection is made between the main reservoir and the brake-pipe through the chamber *C*, the port *G*, and the passage *B*.

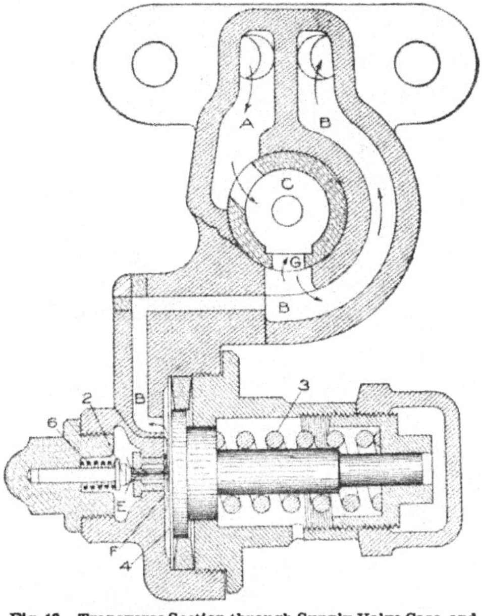

Fig. 13. Transverse Section through Supply-Valve Case, and Central Section through Regulating Valve and Spring-Box of Slide-Valve Feed-Valve.

Air now flowing from the main reservoir will raise the brake-pipe pressure until the pressure in the chamber *F* is sufficient to overcome the tension of the regulating spring (3). This requires 70 pounds in the ordinary equipment. The spring (6) will now seat the regulating valve (2), and cut off communication between the brake-pipe and the chamber *D*. The pressure in the chambers *C* and *D* will then equalize through leakage past the piston (1), when the spring (7) will move the piston (1) and the slide-valve (5) to the left, closing the

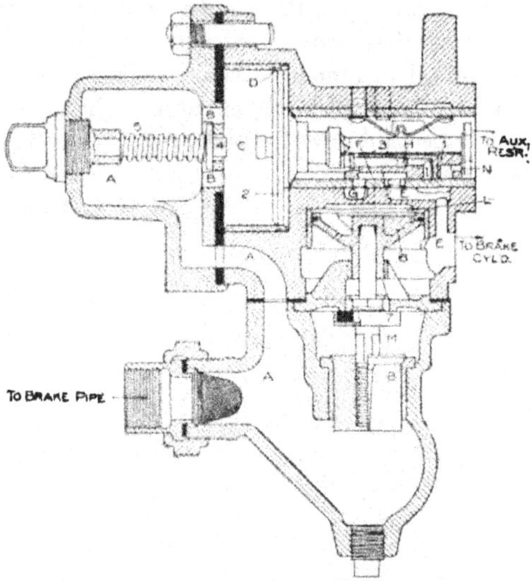

Fig. 14. Quick-Action Triple Valve, Release Position.

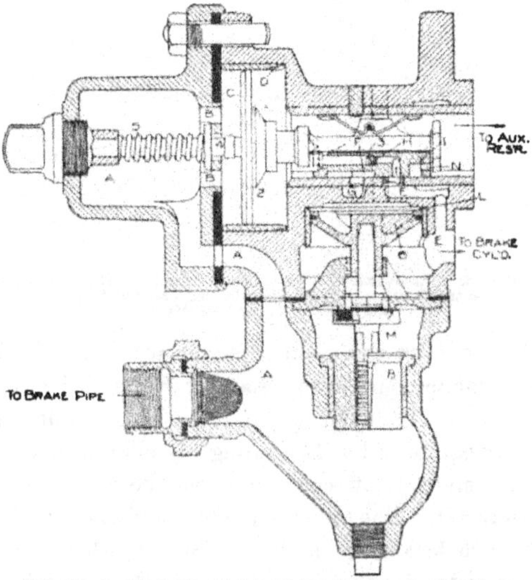

Fig. 15. Quick-Action Triple Valve, Service Application.

port *G* and shutting off communication between the main reservoir and the brake-pipe. A subsequent reduction of brake-pipe pressure will cause the regulating spring (3) to unseat the valve (2), and the accumulated pressure in the chamber *D* will discharge to the brake-pipe. The pressure in the chamber *C* being greater than that in the chamber *D*, the piston (1) will move to the right and uncover port *G*, allowing the brake-pipe to be recharged.

Quick-Action Triple Valve. The quick-action triple valve is shown in its four positions by Figs. 14, 15, 16, and 17. The principal parts are numbered as follows: (1) Slide valve, (2) main piston,

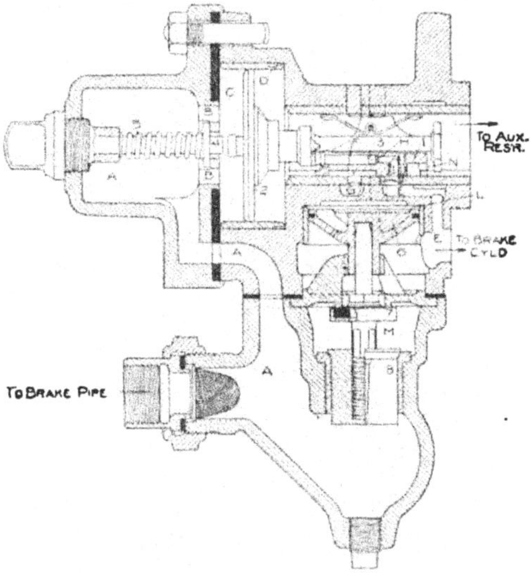

Fig. 16. Quick-Action Triple Valve, Lap Position.

(3) graduating valve, (4) graduating stem, (5) graduating spring, (6) emergency piston, (7) emergency valve, and (8) check-valve.

Charging or Release. Fig. 14 shows the position of the triple valve when the auxiliary reservoir is being charged and the brake is being released. Air enters the triple valve at the point marked "To Brake-Pipe," and follows the passage *A* through the port *B* into the chamber *C*, through the feed-groove *D*, over the slide-valve (1), to the auxiliary reservoir. This flow of air will continue until the pressure in the auxiliary reservoir is equal to the pressure in the brake-pipe. This pressure is 70 pounds in the ordinary equipment. At the same time, air from the brake-cylinder enters the triple valve at the point marked "To Brake-Cyl'd," and passes through the port *E*, through the cavity *F* in the slide-valve (1) (see Fig. 18), and into the exhaust port *G* to the atmosphere.

Service Position. Fig. 15 shows the position of the triple valve during a service application. A reduction of about 5 pounds in the brake-pipe reduces the pressure in the chamber *C*, and causes the piston (2) to move to the left until it strikes the graduating stem (4). This closes the feed-groove *D* and opens the graduating valve (3).

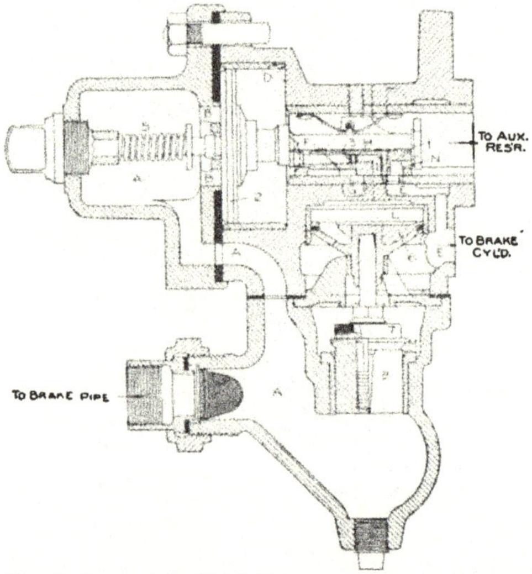

Fig. 17. Quick-Action Triple Valve, Emergency Application.

Auxiliary-reservoir air now flows into the brake-cylinder through the port *H*, passing the graduating valve (3) into the port *I*, which registers with the port *E*.

Lap Position. As soon as the pressure on the auxiliary-reservoir side of the piston (2) has fallen below the brake-pipe pressure, the piston (2) will move to the right and seat the graduating valve (3). This is known as *lap position,* and is shown in Fig. 16. The flow of air from the auxiliary reservoir to the brake-cylinder now ceases. Since the difference in pressure on the two sides of the piston (2) is not sufficient to overcome the frictional resistance of the slide-valve (1), all ports remain in the position shown until another reduction in the brake-pipe is made or until the brake is released. A service reduction of 25 pounds in the brake-pipe will equalize the pressure in the auxiliary reservoir and brake-cylinder at about 50 pounds, this being the maximum pressure obtainable in a service application.

Emergency Application. A sudden reduction of air in the brake-pipe will cause the piston (2) to move to the left with such force that its impingement against the graduating stem (4) will compress the graduating spring (5), as shown in Fig. 17. In this position of the parts, a diagonal slot *J* in the slide valve (1), Fig. 18, registers with the port *K*, which opens into the chamber *L* above the emergency piston (6). This permits auxiliary-reservoir air to act on the piston (6), forcing it down, unseating the emergency valve (7), and allowing the air-pressure in the cavity *M* to enter the brake-cylinder. Brake-pipe pressure then lifts the check-valve (8), and air rushes into the brake-cylinder. At the same time, the auxiliary-reservoir air has a direct passage into the brake-cylinder through the port *N*, which registers with the port *E*. As the opening through the check-valve (8) and the emergency valve (7) is comparatively large, the brake-pipe discharges into the brake-cylinder very rapidly. This causes a quick reduction of pressure in the brake-pipe and affects the next triple in the train, causing it to act in the manner just described. Each triple valve in its turn is affected by the sudden drop in brake-pipe pressure, so that a full emergency application in a 50-car train can be made in about three seconds. The release from an emergency application is made in the same way as the release from a service application. In an emergency application, the pressure in the brake-cylinder rises to about 60 pounds, being about 10 pounds higher than that obtained in a full service application.

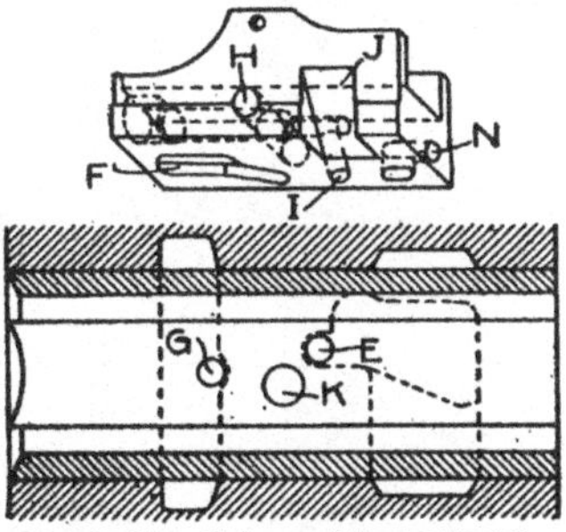

Fig. 18. Slide-Valve and Seat of Quick-Action Triple Valve.

Plain Triple Valve. Fig. 19 shows a section of the plain triple valve. This valve is like the quick-action triple valve, except that the slide-valve (3) and the axis of the main piston (2) are vertical instead of horizontal, and the emergency valve mechanism is omitted. In a service application, the operation of the plain triple is exactly the same as that of the quick-action triple valve, as already described.

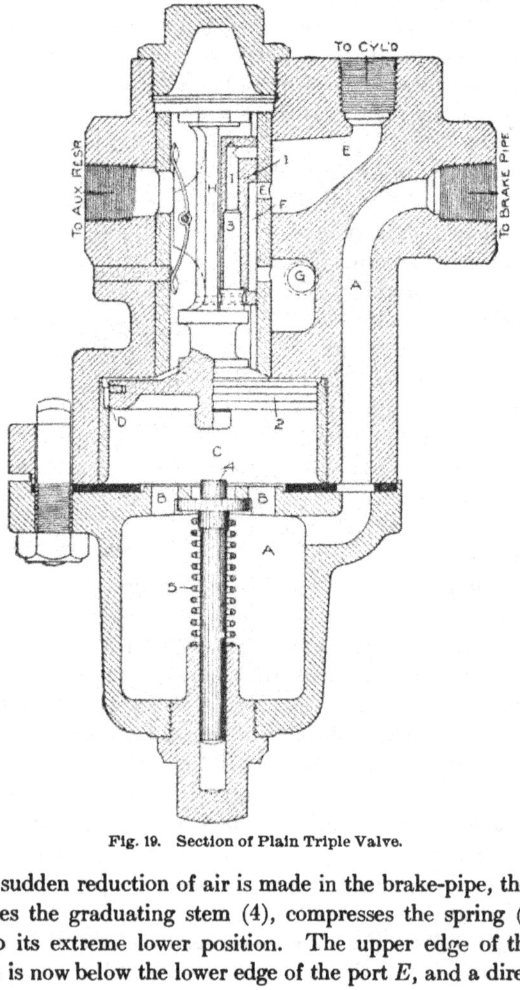

Fig. 19. Section of Plain Triple Valve.

When a sudden reduction of air is made in the brake-pipe, the piston (2) strikes the graduating stem (4), compresses the spring (5), and moves to its extreme lower position. The upper edge of the slide-valve (1) is now below the lower edge of the port E, and a direct communication between the auxiliary reservoir and the brake-cylinder is made. This will cause a quick application of the brake; but the final pressure in the brake-cylinder is no greater than if a full service application were made.

The plain triple valve is used largely on the tender and locomotive equipment, but is gradually being replaced by the quick-action triple.

Combined Freight-Car Cylinder, Reservoir, and Triple Valve. Fig. 20 shows the combined freight-car cylinder and reservoir, which is the usual form of equipment employed on freight-cars. Referring to Fig. 20, (1) is the quick-action triple valve; (2) is the auxiliary reservoir, which is simply a hollow cast-iron shell for the purpose of storing air for use in the brake-cylinder upon the same car; (3) is a release valve usually placed above the auxiliary reservoir for the purpose of releasing the brake in case of necessity; (4) is a pipe connecting

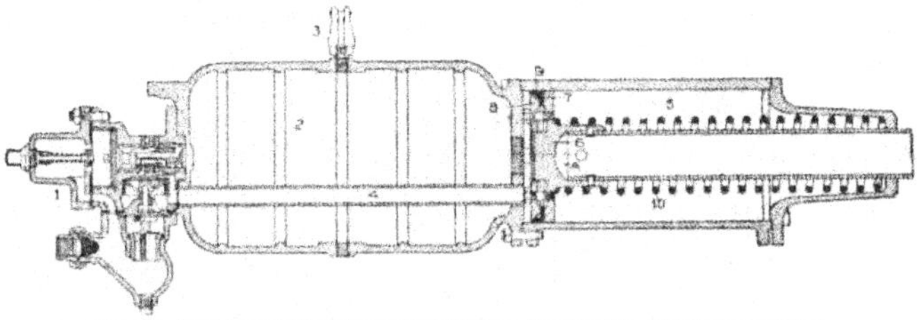

Fig. 20. Combined Freight-Car Cylinder, Reservoir, and Triple Valve.

the triple valve with the brake-cylinder; (5) is the brake-cylinder; (6) is the piston; (7) is the packing leather, which is pressed against the cylinder to prevent air from leaking past the piston; (8) is the follower plate that holds the leather to the piston; (9) is the spring expander, which presses the leather out against the cylinder wall; and (10) is a release spring which brings the piston back to the position shown after the air is exhausted from the cylinder. A rod extends from the valve (3) to either side of the car. If either rod is pulled, the pressure in the auxiliary reservoir will be exhausted and thus will release the brake. There is a small groove called the leakage groove, shown by dotted lines at *A*. This groove permits any small leaks of air which may enter the cylinder, to pass around the piston (6), and thus prevents its being moved outward and setting the brake.

The movement of the piston (6) should be such that the pressure in the auxiliary reservoir and brake-cylinder will equalize at 50 pounds in a full service application. To secure this pressure, the stroke of the piston must be about 8 inches. If the brake is applied when the car is not in motion, the stroke of the piston is called the *standing travel;* when in motion, it is called the *running travel.* Because of the slack in loose fitting parts, shoes pulling down on the wheels, etc., the *running travel* is about 1½ inches greater than the *standing travel.* For this reason, the brake rigging must be adjusted to give a piston

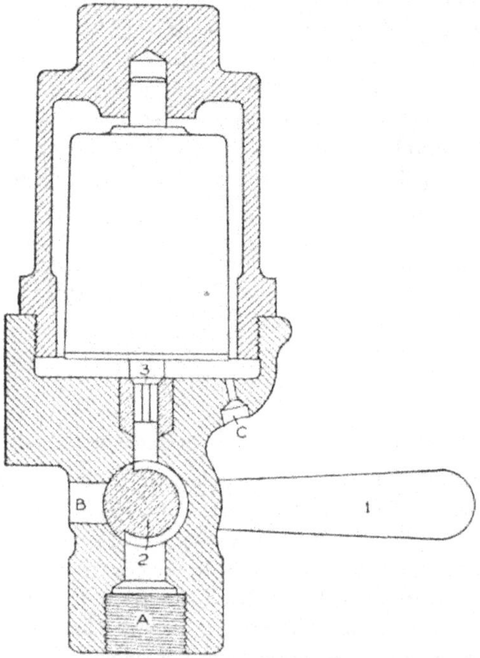

Fig. 21. Section of Pressure-Retaining Valve.

travel of about 6½ inches when the car is not in motion. The brake-cylinder commonly used in freight-car equipment is 8 inches in diameter. When a larger cylinder is used, the auxiliary reservoir must be increased proportionally.

Pressure-Retaining Valve. Fig. 21 shows a section through this valve. A pipe is connected at *A* which comes from the exhaust port of the triple valve. When the valve handle (1) is down, the exhaust from the triple valve enters at *A*, passes the valve (2), and out at the port *B*. If the handle (1) is turned horizontally, as shown in Fig. 21, the air from the triple valve flows around the valve (2), lifting the weighted valve (3), and passes to the atmosphere through the port *C*. A pressure of over 15 pounds will raise the weighted valve (3). When the brake-cylinder pressure has become reduced to 15 pounds, the weighted-valve (3) seats, and the remaining 15 pounds pressure is retained in the brake-cylinder until the handle (1) is turned down. Pressure-retaining valves are used mostly on freight-cars; but some roads that have long, heavy grades use them on passenger cars also. These valves should be so located that they can be reached while the train is in motion. The usual location on freight-cars is at the end of the car, just under the foot-board.

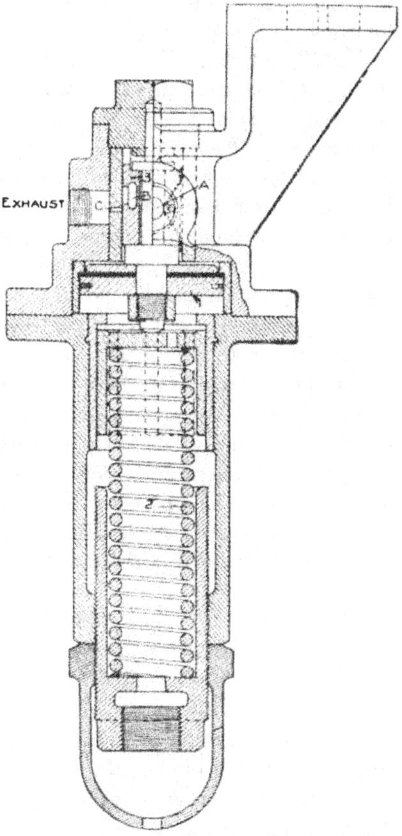

Fig. 22. Section of Automatic Reducing Valve.

High-Speed Brake. It has been known for several years that as the speed of the train is increased, the maximum brake-shoe pressure may also be increased without danger of skidding the wheels. That is, a train going at a speed of 80 miles an hour would require a much greater brake-shoe pressure to skid the wheels than a train going 5 miles an hour. This fact has been taken advantage of in the design of the high-speed brake. Instead of carrying a brake-pipe pressure of 70 pounds when the high-speed brake is in operation, 110 pounds is carried. When a full service application is made, 85 pounds pressure is obtained in the brake-cylinder. If this pressure were allowed to continue in the brake-cylinder until the train stopped, there would be danger of skidding the wheels. In order to prevent this, a valve known as the *automatic reducing valve* is used. This valve is shown in Figs. 22 and 23. The chamber *A* above the piston (1), Fig. 22, has at all times communication with the brake-cylinder by means of a pipe connection. When the pressure in the brake-cylinder is 60 pounds or less, the parts of the valve are in the position shown. If, during a heavy service application, the pressure in the brake-cylinder becomes greater than 60 pounds, its action on piston (1) will be sufficient to overcome the tension in the spring (2); and piston (1), together with the slide-valve (3), will move downward until the port *C* registers with the triangular port *B*, which is always in communication with the chamber *A*. Air from the brake-cylinder now escapes through chamber *A* and ports *B* and *C* to the atmosphere. This exhaust of air will continue until the brake-cylinder pressure is reduced to 60 pounds. The spring (2) then raises the piston (1), causing the slide-valve (3) to close the exhaust port *C*. In the operation just described, the greatest width of the triangular port *B* is exposed to the port *C*. These ports are so proportioned that in this particular position, the surplus air is discharged from the brake-cylinder as rapidly as it is admitted through the service application port of the triple valve.

In an emergency application, the violent admission of air into the brake-cylinder so suddenly increases the pressure that the piston (1) is forced to the lower end of its stroke. In this position, only the apex of the triangular port *B* in the slide-valve (3) registers with the port *C*, and a comparatively slow discharge of brake-cylinder air-pressure takes place while the train is at its highest speed; but the area of the opening of the port *B* gradually increases as the pressure decreases, until the pressure in the brake-cylinder is 60 pounds, after which time the port *C* is closed as in a service application.

The high-speed equipment is used principally on fast passenger trains. Fig. 23 shows the general arrangement under a passenger car of the auxiliary reservoir, brake-cylinder, triple valve, high-speed reducing valve, and pipe connections. The arrangement of the brake-cylinder and the auxiliary reservoir is different from that used on freight-cars, in that they are separate, and the triple valve is bolted to the brake-cylinder instead of to the auxiliary reservoir. On a locomotive used in handling both the high-speed and standard equipments, the engineer's brake-valve is fitted with two feed-valves; and the air-pump, with a duplex governor. One feed-valve is adjusted for 70 pounds, and the other for 110 pounds. One pump-governor is adjusted to maintain a pressure of 90 pounds in the main reservoir, and the other a pressure of 130 pounds.

The equipment on the locomotive may be changed from the standard to the high-speed by closing a cut-out cock in the pipe leading

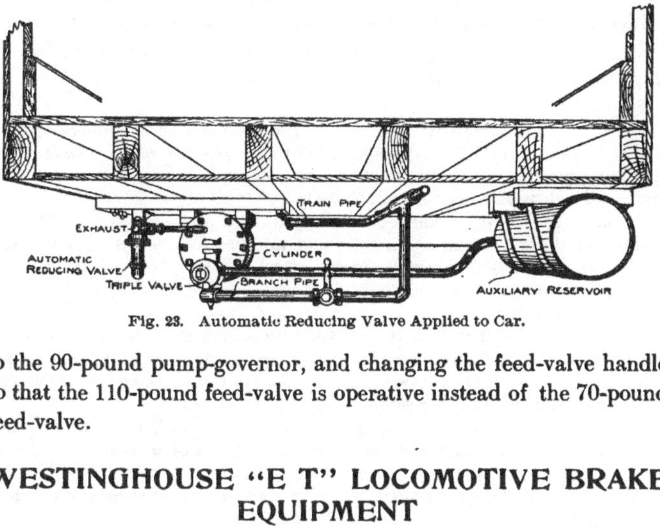

Fig. 23. Automatic Reducing Valve Applied to Car.

to the 90-pound pump-governor, and changing the feed-valve handle so that the 110-pound feed-valve is operative instead of the 70-pound feed-valve.

WESTINGHOUSE "E T" LOCOMOTIVE BRAKE EQUIPMENT

In a paper read before the 1906 meeting of the Air-Brake Association, Mr. F. H. Parke stated that the possible braking power of a single modern locomotive was over 10 per cent of a 50-car freight train, 12 per cent of a 12-car Pullman train, 25 per cent of a 10-car passenger train, and 35 per cent of a 6-car passenger train. These figures show that the brake equipment of the locomotive should receive special attention, and that the brake equipment commonly used on locomotives should be improved.

The first step taken in this direction was the development of the combined automatic and straight-air equipment for locomotives. This system provided a means for applying and releasing the brakes on the cars in the train. It greatly increased the control of the engine in switching and in handling the slack in long freight trains, but it had many undesirable features. It was greatly simplified and improved by the invention of the so-called "E T" locomotive brake equipment.

The "E T" equipment possesses all the advantages of the combined automatic and straight-air equipment, and several additional ones which are necessary to give satisfactory results in braking long trains. It can be applied to any locomotive without change or modification of any of its parts, and the locomotive so equipped can be used for any class of service. The Westinghouse Company makes the following claims for the "E T" equipment:

1. "The locomotive brakes may be controlled with or independently of the train brakes, and this without regard to the position of the locomotive in the train, whether coupled to another, as in double heading, or used as a helper and assigned to any position in the train.

2. "They may be applied with any desired pressure between the minimum and the maximum attainable; and this pressure will be automatically maintained in the locomotive brake-cylinders, regardless of leakage and variation in piston travel—undesirable though these defects are—until released by the brake-valve.

3. "They will remain applied when the engineer places the automatic brake-valve handle in full release, then in lap position preparatory to making the second application in a two-application stop, thus making a more uniform stop and requiring a lighter second application.

4. "They can be perfectly graduated on or off, either in the automatic or in the independent application; hence, in all kinds of service, the train may be handled without shock or danger of parting; and in passenger service especially, smooth and accurate stops can be made with greater ease than was heretofore possible."

Fig. 24 gives the names of all parts used in the equipment, and shows the scheme and sizes of piping. The parts not to be found in the quick-action automatic equipment as commonly installed on the locomotive, are as follows:

1. A *duplex pump-governor*.

2. A *distributing valve*, and small double-chamber reservoir to which it is attached. This valve performs the functions of triple valve, auxiliary reservoir, double check-valve, and high-speed reducing valve on the locomotive.

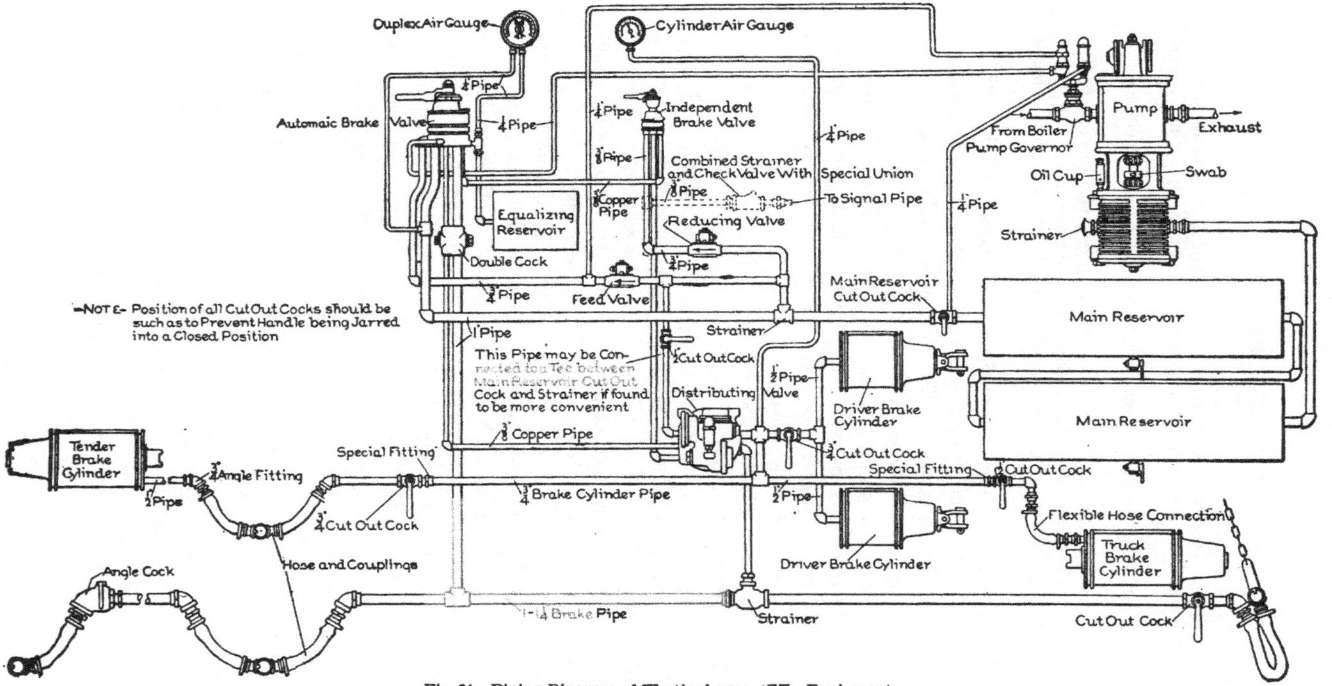

Fig. 24. Piping Diagram of Westinghouse 'ET Equipment.

3. *Two brake-valves*, one of which is automatic and operates both the train and locomotive brakes; the other an independent valve operating the locomotive brakes only.

4. A *feed-valve*, located in the reservoir pipe to regulate the brake-pipe pressure.

5. A *reducing valve*, which reduces the main-reservoir pressure for the independent brake-valve and for the air-signal system when used.

6. A *single-pointer gauge* to indicate the locomotive brake-cylinder pressure.

Manipulation. The instructions for operating the "ET" equipment are very similar to those given for the combined automatic and straight-air equipment. The automatic brake-valve has six fixed positions for its handle, while the independent brake-valve has but five. The positions for the automatic brake-valve are: *Release, running, holding* (driver brake), *lap, service*, and *emergency;* and those for the independent brake-valve are: *Release, running, lap, slow application*, and *quick application.*

The handles of both brake-valves should be kept in running position when not in use.

To make a service application on both locomotive and train brakes, move the handle of the automatic brake-valve to service position long enough to secure the required brake-pipe reduction; then move the handle back to lap position. All brakes will remain applied as long as the valve remains in this position.

The train brakes may be released by moving the handle of the automatic brake-valve to release position. When the valve is in release position, care should be exercised that the brake-pipe does not become overcharged. This action does not release the locomotive brakes. If the handle is moved from release to holding position, the locomotive brakes will still remain applied. They may be graduated off by short successive movements of the handle between running and holding positions, or by placing the handle at once in the running position. If a full stop is not desired, the handle of the automatic brake-valve should be placed in service position until the required reduction in brake-pipe pressure is obtained; then moved to lap position. After the speed has dropped sufficiently, place the handle in release position until all the train brakes are released and the slack has had an opportunity to adjust itself; then place the handle in running position to release the locomotive brakes.

An emergency application is made with the automatic brake-valve in exactly the same manner as with the engineer's brake-valve commonly installed on locomotives.

If only the independent brake-valve is being used, the handle of the automatic brake-valve should be carried in running position. The locomotive brakes may be released by placing the handle of the independent brake-valve in running position. When the handle of the automatic brake-valve is not in running position, the only way in which the locomotive brakes can be released is by placing the handle of the independent brake-valve in release position. The locomotive brakes may be released under any and all conditions by placing the handle of the independent brake-valve in release position. The independent brake-valve should be used very carefully when handling long trains or in switching service, as damage to draft gears might result if the slack in the train is permitted to run out hard. If an emergency case arises, the automatic brake should be applied instantly, even though the independent brake is being used. The safety-valve on the distributing valve will prevent any excessive brake-cylinder pressure on the locomotive.

In handling trains on long grades, the application of the train brakes and locomotive brakes should be alternated to prevent any overheating of the wheels. When leaving the locomotive while doing work about or near it, the independent brake-valve should always be left in application position.

When double-heading, the double cut-out cock in the brake-pipe below the automatic brake-valve should be closed, and the valve should be left in lap position, except on the locomotive from which the brakes are being operated.

In order to simplify the description of the various parts of the equipment, the following names of pipes are given (see Fig. 24):

Reservoir Pipe. The pipe connecting the main reservoir to the automatic brake-valve, distributing valve, feed-valve, and reducing valve.

Feed-Valve Pipe. The pipe connecting the feed-valve to the automatic brake-valve.

Reducing-Valve Pipe. The pipe connecting the reducing valve to the independent brake-valve and the air-signal system.

Brake-Pipe. The pipe connecting the automatic brake-valve to the distributing valve and the triple valves on the cars in the train.

Brake-Cylinder Pipe. The pipe connecting the distributing valve to the brake-cylinders on the engine and tender.

Application-Chamber Pipe. The pipe connecting the application chamber of the distributing valve to the automatic brake-valve, through the independent brake-valve.

Double-Heading Pipe. The pipe connecting the application chamber of the distributing valve to the automatic brake-valve, through the double cut-out cock.

The main-reservoir cut-out cock is used to cut off the supply of air when removing any part of the equipment for cleaning or repairing, except the governor. Before it is closed, however, the double cut-out cock below the automatic brake-valve should be turned to close the brake-pipe, and the handle of the automatic brake-valve should be placed in release position. This is done so as to prevent lifting from its seat the rotary valve of the automatic brake-valve or the slide-valve of the feed-valve. The automatic brake-valve receives air from the main reservoir direct and through the feed-valve. The check-valve in the signal pipe prevents air from flowing back from the signal pipe when the independent brake-valve is being used. The pump-governor has three pipe connections—one from the reservoir pipe to the maximum pressure head; one from the feed-valve pipe to the upper connection of the excess pressure head; and one from the automatic brake-valve to the lower connection of the excess pressure head.

Fig. 25. Distributing Valve and Double-Chamber Reservoir of "ET" Equipment.

Distributing Valve. The distributing valve, shown in Figs. 25 and 26, has five pipe connections. On the reservoir are cast the following letters indicating these connections: *MR*—main reservoir pipe; *DH*—double-heading pipe; *AC*—application-chamber pipe; *BC*—brake-cylinder pipe; and *BP*—brake-pipe.

Fig. 27 is a section through the distributing valve. The prin-

Fig. 26. Part Sectional View of Distributing Valve and Double-Chamber Reservoir of "ET" Equipment.

ciples governing the operation of this valve are the same as those previously described in the standard equipment commonly used. The chief difference is the manner in which air-pressure is supplied

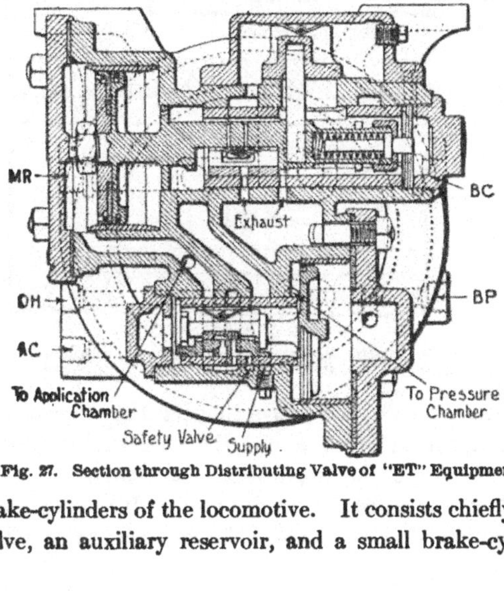

Fig. 27. Section through Distributing Valve of "ET" Equipment.

to the brake-cylinders of the locomotive. It consists chiefly of a plain triple valve, an auxiliary reservoir, and a small brake-cylinder, the piston-rod of which operates two slide-valves instead of being connected to the brake-rigging. The piston travel is always constant. One of the slide-valves mentioned admits air to the brake-cylinders of the locomotive, and the other releases or exhausts this air. The triple valve and auxiliary reservoir are used in automatic applications only, and are called, respectively, the *equalizing portion* and the

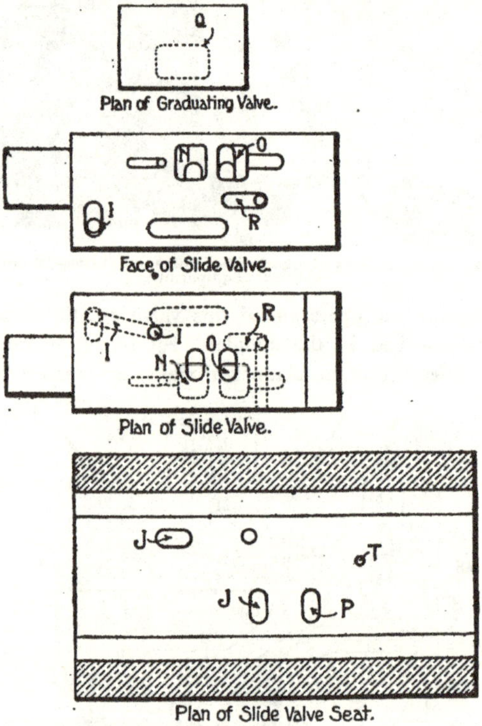

Fig. 28. Graduating Valve, Equalizing Slide-Valve, and Slide-Valve Seat of Distributing Valve of "ET" Equipment.

pressure chamber. The slide-valve connected to the piston-rod of the small brake-cylinder, which admits air to the brake-cylinders of the locomotive, is called the *application valve,* while that one which exhausts this air is called the *exhaust valve.* It is easily seen that the entire operation of the locomotive brakes consists in admitting or releasing air-pressure into or out of the application chamber. In independent applications, this is done directly by operating the independent brake-valve; while in automatic applications, it is accomplished by means of the equalizing piston and the air-pressure stored in the pressure chamber.

Fig. 28 is given to show the correct location of ports in the equalizing valve and slide-valve.

Since the ports in the valve cannot be clearly indicated in Fig. 27, diagrammatic illustrations shown in Figs. 29, 30, 31, 32, 33, 34,

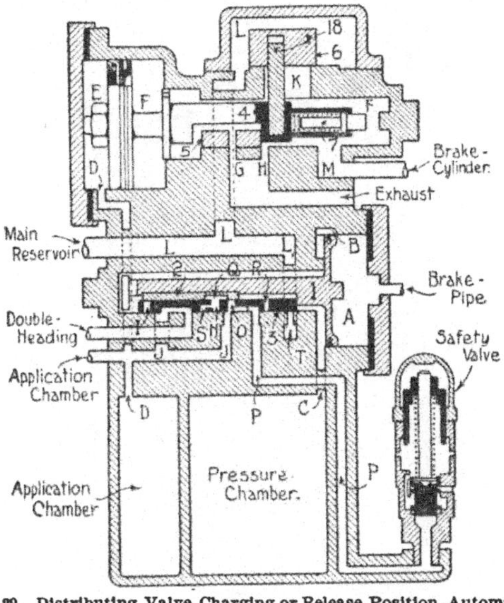

Fig. 29. Distributing Valve, Charging or Release Position, Automatic or Independent.

35, and 36 will be referred to in explaining the operation of the valve. These diagrams show the parts distorted and not as actually constructed.

The operation of the valve when automatic applications are made, is as follows:

Charging. Fig. 29 shows the movable parts of the valve in charging position. In this position, the chamber *A* is in connection with the brake-pipe; and air is free to pass around the top of the piston

(1) through the feed-groove *B* and the port *C*, to the pressure chamber, until the pressure on both sides of the piston becomes equal.

Release. The position shown in Fig. 29 is also the release position, and is the position the parts take when the automatic brake-valve handle is placed in release position. In this position, the pressure in the chamber *A* is greater than that in the application chamber; consequently the equalizing piston (1) is moved to the position shown. This movement of the piston moves the graduating valve (2) and the slide-valve (3) to the release position shown, but does not release the

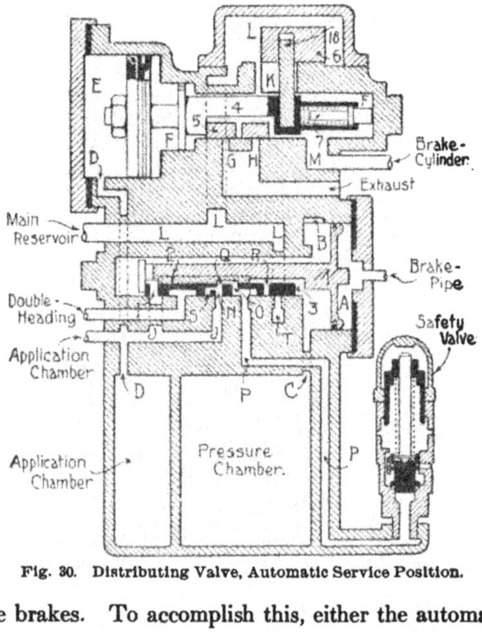

Fig. 30. Distributing Valve, Automatic Service Position.

locomotive brakes. To accomplish this, either the automatic brake-valve must be placed in running position or the independent brake-valve must be moved to release position. In either case, the application-chamber pipe is opened to the atmosphere, and the air in the application chamber is exhausted. The air in the chamber *E* will also be exhausted, since it is connected to the application chamber by the port *D*. This permits the brake-cylinder pressure in the chamber *F* to move the piston (4) to the left until the exhaust ports *G* and *H* permit the brake-cylinder pressure to escape. The double-heading pipe must always be kept closed at the double cut-out cock below the automatic brake-valve, unless there are two engines at the head of the train. In this case, the engine from which the brakes are controlled should have its double-heading pipe closed, while on the other engine it should be open.

Service. When a service application is made with the automatic brake-valve, the brake-pipe pressure in the chamber *A* is reduced; and piston (1), together with the graduating valve (2) and the slide-

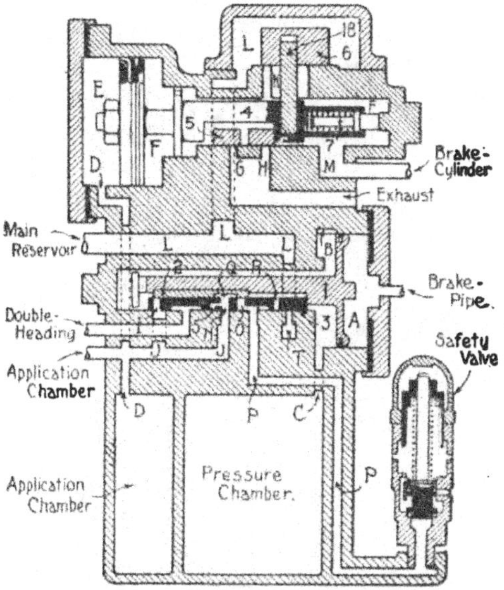

Fig. 31. Distributing Valve.—Service Lap Position.

valve (3), is moved toward the right to the position shown in Fig. 30. In this position, the port *I* in the slide-valve registers with the port *J* in the seat, and permits air from the pressure chamber to flow into the application chamber and the chamber *E* through the port *D*. This pressure forces the application piston (4) to the right, causing the exhaust valve (5) to close the exhaust ports *G* and *H*, and the application valve (6) to uncover the port *K;* also causing the graduating spring on the stem (7) to be compressed. Air from the main reservoir is now free to flow from the chamber *L* through the port *K* and passage *M* to the brake-cylinders.

In the movement just described, the ports *N* and *O* in the slide-valve register with the ports *J* and *P* in the seat, and are connected by the cavity *Q* in the graduating valve. This connects the application chamber with the safety-valve, which, being adjusted to open at 53 pounds, limits the cylinder pressure to this amount during a full service application.

Service Lap. If the brake-pipe reduction is not sufficient to

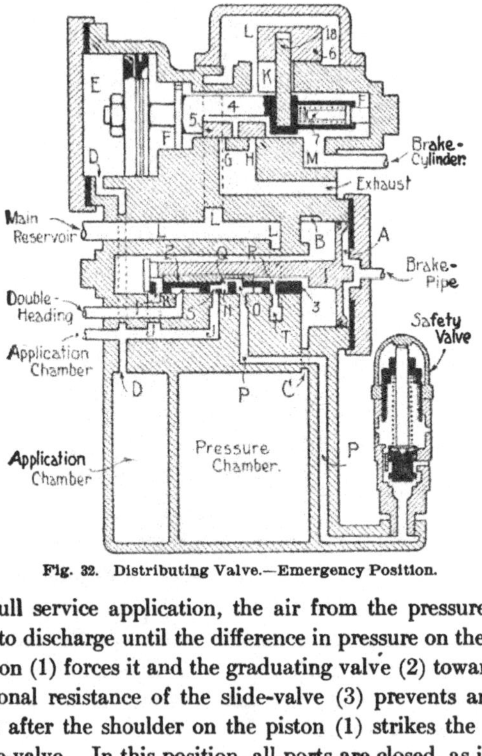

Fig. 32. Distributing Valve.—Emergency Position.

cause a full service application, the air from the pressure chamber continues to discharge until the difference in pressure on the two sides of the piston (1) forces it and the graduating valve (2) toward the left. The frictional resistance of the slide-valve (3) prevents any further movement after the shoulder on the piston (1) strikes the right end of the slide-valve. In this position, all ports are closed, as in Fig. 31, and the valve is in *service lap*. Air continues to flow through the port *K* and the passage *M* into the brake-cylinders, until their pressure is slightly in excess of that in the application chamber. This difference in pressure on the two sides of the piston (4), assisted by the graduating spring on the stem (7), forces the piston (4) to the position shown in Fig. 31. This movement of the piston (4) results in application valve (6) closing the port *K*, but does not move the exhaust valve (5). The brake-cylinder pressure is then about the same as that in the application chamber.

Emergency. When a sudden and heavy reduction of air-pressure is made in the brake-pipe, the piston (1) is forced to the right by the pressure in the pressure chamber until it strikes the gasket as shown in Fig. 32. This movement causes the slide-valve (3) to uncover the port *J*; and air from the pressure chamber passes quickly into the application chamber and becomes equalized. When the automatic brake-valve is placed in emergency position, the ports in the valve connect the equalizing reservoir to the application-chamber pipe. Air from the equalizing reservoir then passes into the application chamber, and, with that from the pressure chamber, equalizes at about 60 pounds. Air from the main reservoir enters the slide-valve chamber through the pipe *L* and the ports *T* and *R*, and passes into the pressure and application chambers. Air now escapes from the application chamber through the port *J* into the cavity *S*, through a small port into the port *N*, and thence out through the safety-valve. Air escapes through the safety-valve more rapidly than it can be supplied through the ports *R* and *T*, and thus prevents the pressure from becoming higher than is desired.

In high-speed service, the feed-valve is set to maintain a brake-pipe pressure of 110 pounds instead of 70; and a main-reservoir pressure of 130 or 140 pounds is carried. The pressure in the application chamber, under these conditions, is increased to about 85 pounds; but air escapes through the cavity *S* and port *N* at about the same rate as in the high-speed reducing valve, until the pressure is only about 60 pounds. The pressure in the application chamber does not drop below about 60 pounds, because, under these conditions, air from the main reservoir is supplied through the ports *R* and *T* faster than it can escape through the restricted passages to the safety-valve.

Emergency Lap. In emergency applications, the process above described continues until the brake-cylinder pressure slightly exceeds the pressure in the application chamber, when all parts move back to emergency lap position, as shown in Fig. 33. Release is accomplished in the same manner as described under Fig. 29.

In operating the locomotive brakes with the independent brake-valve, the action of the distributing valve is as follows:

Independent Application. When making an application, the equalizing piston (1) occupies the same position as shown in Fig. 34. Air is admitted into the application chamber from the main reservoir through the reducing valve, at 45 pounds' pressure. This pressure

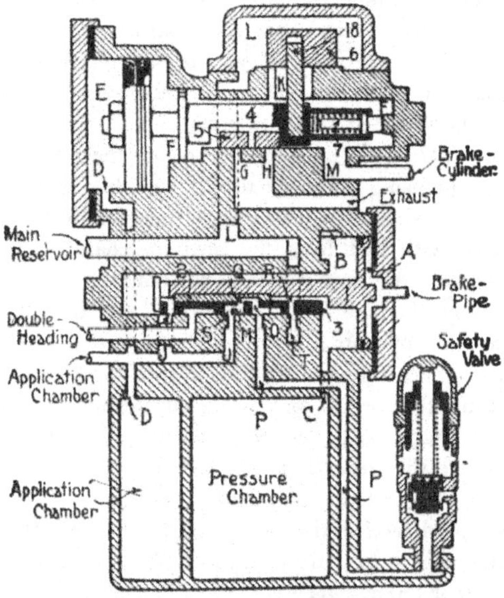

Fig. 33. Distributing Valve.—Emergency Lap Position.

also exists in the chamber *E*, and forces the piston (4) to the right, as shown. This movement causes the application valve (6) to uncover the port *K*, and air from the main reservoir passes through the passage *M* into the brake-cylinders. Air continues to flow into the brake-cylinders until their pressure and that in the chamber *F* slightly exceeds that in the chamber *E*, when the piston (4) will be moved to the left, causing the application valve (6) to close the port. This position, shown in Fig. 35, is known as *independent lap*.

It is easily seen that the action of the piston (4) will always maintain about the same pressure in the brake-cylinders as exists in the application chamber.

Independent Release. If the handle of the independent brake-valve is placed in release position, the air in the application chamber escapes directly to the atmosphere. This permits the brake-cylinder pressure in the chamber *F* to force the piston (4) to the left, causing the application valve (6) to close the port *K*, and the exhaust valve (5) to open the ports *G* and *H*, as shown in Fig. 29. Air is now free to escape from the brake-cylinders until the valve is placed in lap

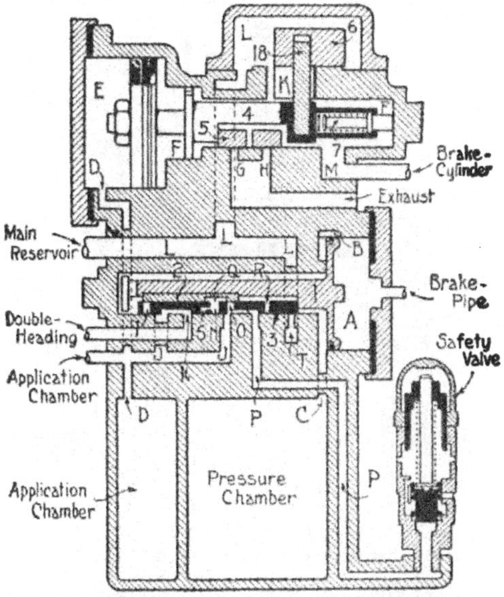

Fig. 34. Independent Application of Distributing Valve.

position or until the brake-cylinders are entirely exhausted. If the handle of the independent brake-valve is placed in lap position before all the air is exhausted from the brake-cylinders, the parts of the distributing valve will move to independent lap position, as shown in Fig. 35. In this way, the independent release may be graduated as desired.

Safety-Valve. One of the essential parts of the distributing valve is the safety-valve. The principle of its action is shown in the section given in Fig. 36. Its construction is such as to cause it to close quickly with a *pop* action, which insures a firm seating. The spring should be adjusted so that the valve will open at 53 pounds. This is accomplished by removing the cap nut (1) and screwing, up or down, an adjusting nut (2).

Automatic Brake-Valve. The automatic brake-valve not only performs the functions of the standard engineer's valve commonly

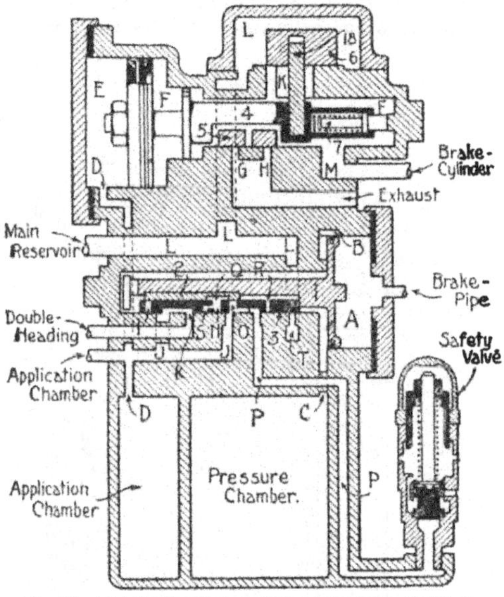

Fig. 35. Distributing Valve.—Independent Lap Position.

installed on locomotives, but also those necessary to obtain all the desirable features of the distributing valve.

Fig. 37 is taken from a photograph of this valve, with the handle in running position.

Fig. 38 is a top view, showing the six positions of the brake-valve handle.

Fig. 39 shows two views, the upper one being a section through the rotary-valve chamber, the rotary valve being removed; and the lower one, a vertical section of the entire valve. A plan of the top view of the rotary valve is shown on the left.

The description of the operation of the valve in its different positions, will be given in the order in which it is most generally used.

Charging or Release Position. In this position, air flows directly from the main-reservoir pipe, through the port *A* in the rotary valve and the port *B* in the valve-seat, into the brake-pipe. This quickly recharges the train-brake system and releases the train brakes, but does not release the locomotive brakes, if they are applied. The port *C* now registers with the port *D*, and permits main-reservoir pressure to enter the chamber *E*, and acts on the equalizing piston (2), forcing it downward and closing the discharge valve. In this position, the port *F* in the rotary valve (1) registers with the warning port *G* in the valve-seat, permitting a small amount of air to escape into the exhaust cavity *H*. This serves to make enough noise to attract the engineer and notify him that the valve still remains in release position. A small groove in the face of the rotary valve connects the port *F* with the port *I*, and permits main-reservoir pressure to act on the excess-pressure head of the pump-governor. If the handle of the automatic brake-valve is permitted to remain in this position too long, the brake-pipe and auxiliary reservoirs would become charged to main-reservoir pressure. The handle should be moved to *running* or *holding* position before this occurs.

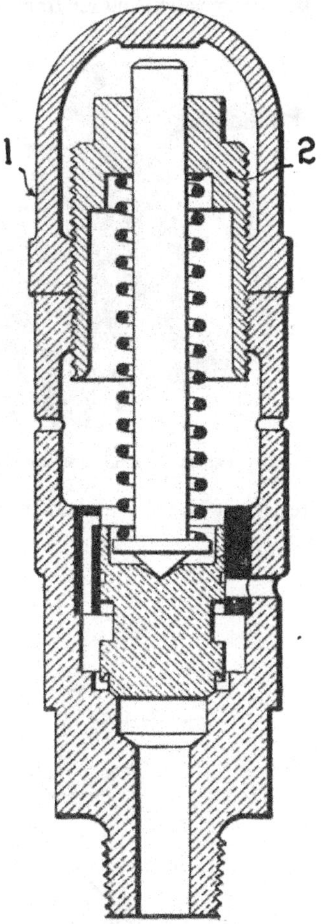

Fig. 36. Section of Safety-Valve of Distributing Valve.

Running Position. In running position, all the train and locomotive brakes are released, and the auxiliary reservoirs are charged. The ports *J* and *B* in the valve-seat are connected by the cavity *K* in the rotary valve; and air from the feed-valve pipe passes directly into the brake-pipe and recharges the auxiliary reservoirs. The air in the brake-pipe will not attain a pressure greater than that for which the feed-valve is set. The ports *L* and *M* in the valve-seat are connected by the cavity *N* in the rotary valve; and the pressure on the equalizing piston (2) and in the equalizing reservoir is the same as that in the brake-pipe. The port *F* in the rotary valve registers with the port *I* in the valve-seat, and permits main-reservoir

Fig. 37. Automatic Brake-Valve.

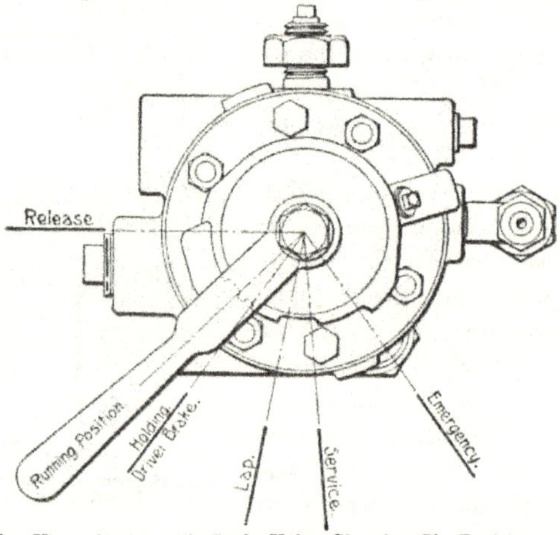

Fig. 38. Top View of Automatic Brake-Valve, Showing Six Positions of Handle.

pressure to pass to the excess-pressure head of the pump-governor. The port *O* in the rotary valve registers with the port *P* in the valve-

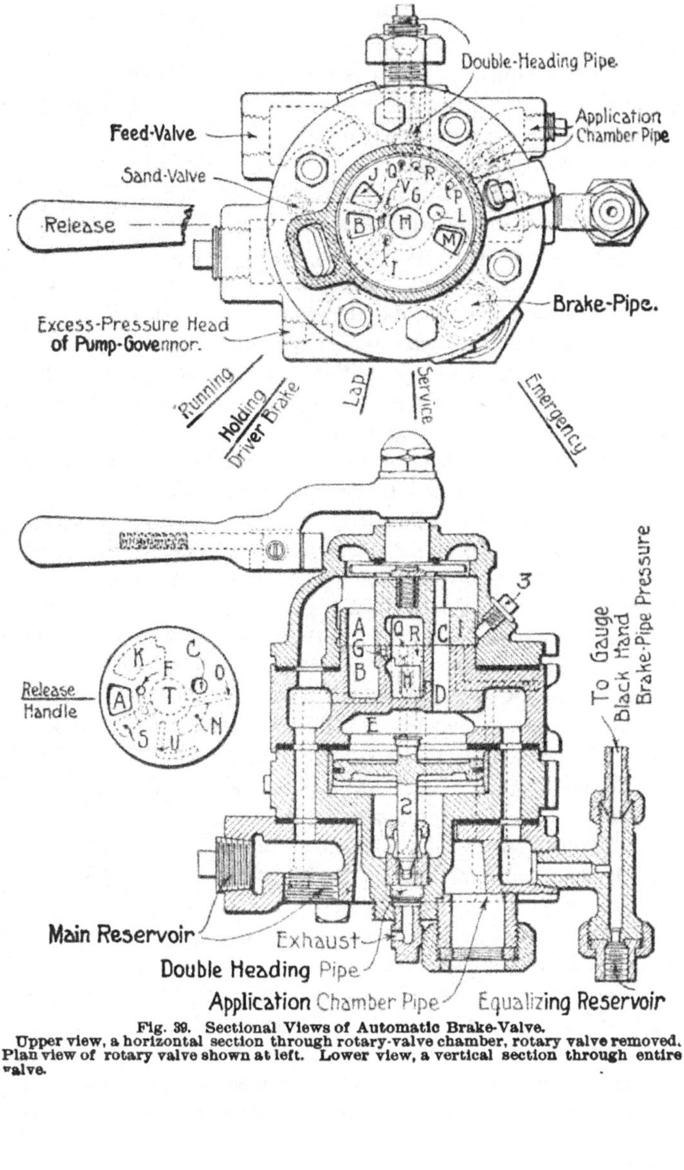

Fig. 39. Sectional Views of Automatic Brake-Valve.
Upper view, a horizontal section through rotary-valve chamber, rotary valve removed. Plan view of rotary valve shown at left. Lower view, a vertical section through entire valve.

seat, and connects the application-chamber pipe with the exhaust cavity *H*.

Service Position. In this position, the brake-pipe pressure is gradually reduced and causes a service application. The port *O* in the rotary valve registers with *Q* in the valve-seat, and permits air to discharge from the chamber *E* and the equalizing reservoir into the exhaust chamber *H*. The port *Q* is restricted, and causes a gradual discharge of air from the equalizing reservoir. As the pressure above the equalizing piston (2) is reduced, the brake-pipe pressure below forces the piston (2) upward, opening the discharge valve and exhausting air from the brake-pipe into the atmosphere. When the pressure in the chamber *E* is reduced the required amount, the handle of the brake-valve is moved to lap position. Air will continue to exhaust from the brake-pipe through the discharge valve, until the pressure below the piston (2) is slightly less than that above. Equalizing piston (2) will then be forced downward, closing the equalizing valve. By this process, it will be seen that the reduction of the pressure in the equalizing reservoir determines that in the brake-pipe.

Lap Position. This is the position the valve occupies while holding the brakes applied, to prevent loss of air from the main reservoir in case of a break-in-two, and when another engine in the train is handling the brakes. All ports are closed except the port *O* in the rotary valve, which connects with the port *R* in the valve-seat. In double-heading, these ports connect with the application chamber in the distributing valve, and permit the air to exhaust into the atmosphere when the automatic brakes are being released.

Release Position. The action of the valve in this position has been described under *charging* or *release*.

Holding Position. In this position, all train brakes are released, but the locomotive brakes are held applied. The only difference between the *running* and *holding* positions is that in the former the application chamber of the distributing valve is open to the atmosphere, while in the latter it is not.

Emergency Position. In this position, the port *S* in the rotary valve registers with the port *M* in the seat, and air discharges from the brake-pipe through the cavity *T*, into the exhaust chamber *H*. These ports are proportioned in such a manner that a large volume of air is suddenly discharged from the brake-pipe, causing all triple valves and the distributing valve to go to the emergency position. The cavity U in the rotary valve registers with L and P in the valve-seat, and permits the air from the equalizing reservoir to flow into the application chamber of the distributing valve. The ports C and V register, and allow air from the main reservoir to flow to the sand valve, thus applying sand to the rails.

Plug 3, shown in Fig. 39, is placed in the top of the case at a point to fix the level of an oil bath in which the rotary valve operates.

Independent Brake-Valve. The independent brake-valve is of the rotary-valve type. Fig. 40 is taken from a photograph of the valve. The general construction of the valve is represented in Fig. 41. The lower view shows a vertical section of the entire valve, with a top view of the rotary valve on the right; while the upper one shows a horizontal section taken through the valve body with the rotary valve removed. All pipe connections, and the different positions of the handle, are shown.

Fig. 40. Independent Brake-Valve.

The action of the valve when placed in the different positions is as follows:

Running Position. When the independent brake is not in use, the independent brake-valve should always be carried in this position. The ports A and B in the valve-seat are connected by the port C in the rotary valve (1). This establishes communication between the application chamber of the distributing valve and the port P (see Fig. 39) of the automatic brake-valve, so that the former can be operated by the latter. If the independent brakes are being operated with the automatic brake-valve in running position, they can be released by simply moving the independent brake-valve to running position, since in this position the air in the application chamber of the distributing valve can escape through the automatic brake-valve.

Service Position. In this position, the ports D and B in the valve-seat are connected by the groove E in the rotary valve, allowing air to flow from the main reservoir to the application chamber. The air-supply from the main reservoir is reduced by the reducing valve to 45 pounds. This is the maximum pressure that can be obtained in the brake-cylinders when using the independent brake-valve.

Lap Position. This position is used to hold the locomotive brakes after having been applied by using the independent brake-valve. All operating ports are closed.

Release Position. In this position, the locomotive brakes will be released when the automatic brake-valve is not in running position.

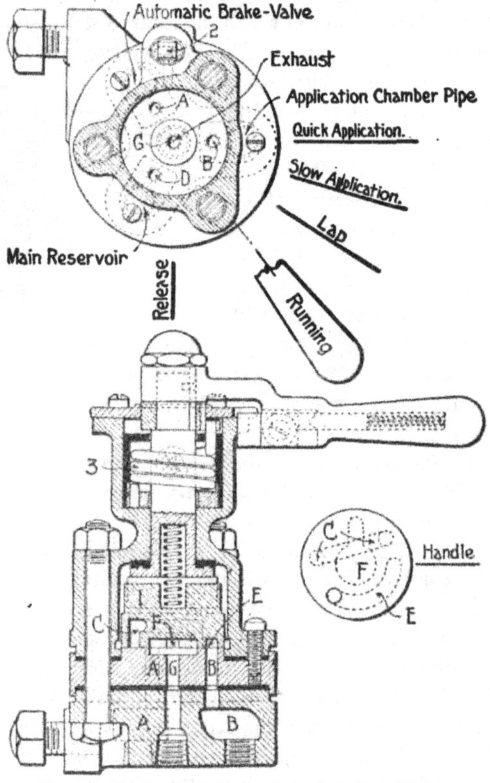

Fig. 41. Sectional Views of Independent Brake-Valve.
Upper view, a horizontal section through valve body, rotary valve removed. Lower view, a vertical section through entire valve. Plan view of rotary valve shown at right.

The port *B* in the valve-seat registers with the cavity *F* in the rotary valve, and air from the application chamber of the distributing valve exhausts into the atmosphere.

If the valve is left in this position, it is impossible to operate the locomotive brakes by means of the automatic brake-valve. For this reason, the coil spring (3) is provided, which always returns the handle to running position as soon as the engineer lets go of it. The purpose of the oil plug (2) is the same as that described in connection with the automatic brake-valve.

Reducing Valve. This is shown in Fig. 42, and is almost identical with the feed-valve. The only difference in their construction is in the manner of adjustment. The principle of its action has already been described.

Pump-Governor. The pump-governor is shown in Fig. 43, with its different pipe connections named. When the automatic brake-valve is in release, running, or holding position, air from the main reservoir flows through the automatic brake-valve into the chamber A below the diaphragm (1). Air from the feed-valve enters above the diaphragm (1), assisting the spring (2) to hold it down. Since the spring (2) is adjusted to a compression of 20 pounds, the diaphragm (1) will not be lifted until the main-reservoir pressure exceeds the feed-valve pipe pressure by this amount. When this occurs, the diaphragm (1) is lifted, and the pin valve is opened. This permits main-reservoir pressure to act on the piston (3), forcing it downward and practically stopping the pump. When the main-reservoir pressure in the chamber A becomes slightly reduced, the diaphragm (1) is forced downward and the pin valve is closed. The air confined above the piston (3) escapes through the port B; the piston (3) is lifted by the action of the spring (4); and the pump starts working. When the automatic brake-valve is in any position other than release, running, or holding, the port connecting the automatic brake-valve with the chamber A is closed, and this governor head is cut out of action. The pump is then controlled by the other governor head, which is always connected with the main reservoir. Its action is similar to that just described. Both governor heads are adjusted by

Fig. 42. Reducing Valve.

screwing up or down on adjusting plugs (5). As both governor heads have a small vent port *B* from which air escapes whenever pressure is present above the piston (3), one of these should be plugged to avoid a waste of air. A small port in the valve (6) permits steam to

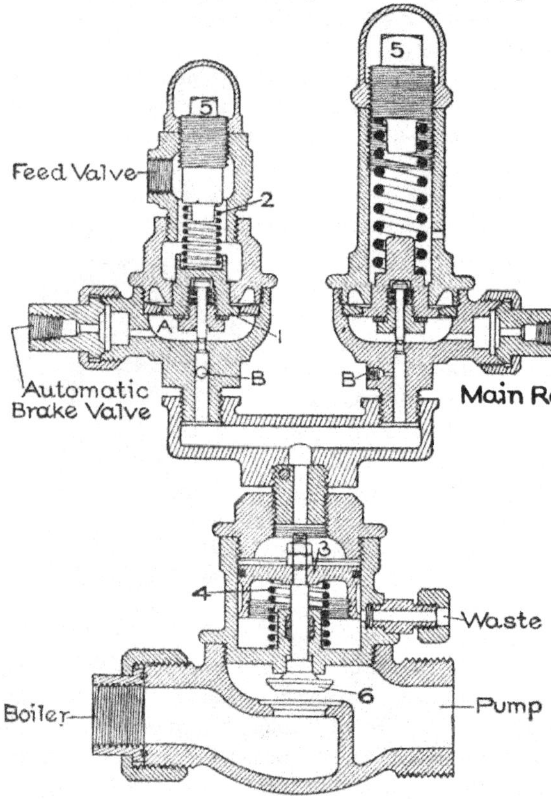

Fig. 43. Vertical Section through Pump-Governor.

enter the pump when it is cut out of action by the governor, which prevents freezing in cold climates.

WESTINGHOUSE TYPE "K" TRIPLE VALVE

The standard form of quick-action triple valve commonly used in freight and passenger service, has until recently proven very satisfactory. In the last few years, however, with heavier locomotives capable of handling 100-car trains fitted with air-brake equipment, they have failed to meet all the requirements. Realizing the changed conditions and the importance of meeting them, the Westinghouse Company has recently perfected the "K" triple valve.

Some of the undesirable features of the standard quick-action triple which the "K" triple overcomes, are as follows:

(*a*) The failure of a portion of the brakes in a long train to apply.

(*b*) A complete release of the brakes at the forward end of the train before the brake-pipe pressure which has brought this about can reach the triple valves near the end of the train. This action permits the slack to run out hard, and creates excessive strains on the draft gears, often resulting in a break-in-two.

(*c*) Overcharging the auxiliary reservoirs at the forward end of the train while releasing the brakes. The result of this action is a re-application of the forward brakes when the brake-valve handle is placed in running position.

The outward appearance of the "K" triple valve, when attached to the auxiliary reservoir, is so much like the standard quick-action

Fig. 44. Westinghouse Type "K" Freight Triple Valve.

triple that a thin web is cast on the top part of the body as a distinguishing mark. The designating mark "K-1" or "K-2" is also cast on the side of the body. The "K" triple is made in two sizes—the "K-1" for use with the 8-inch freight-car brake-cylinder; and the "K-2," with the 10-inch freight-car brake-cylinder (see Fig. 44).

This new valve embodies every feature possessed by the standard quick-action triple, and three additional ones—namely, the *quick service, retarded release,* and *uniform recharge.* It operates in perfect harmony with the standard triple, and often improves the action of the latter when the valves are mixed in the same train. The two types of valves have many parts in common and are interchangeable. The standard triple may be transformed into the "K" triple by preserving all of the old parts, save the body, slide-valve, bush, and

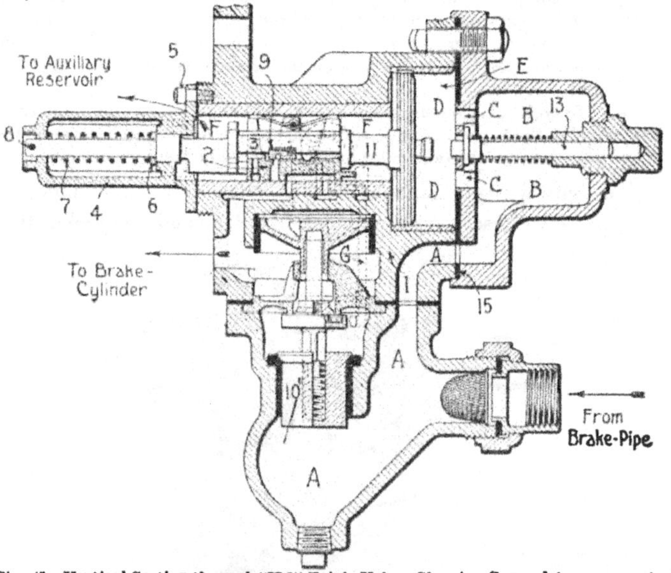

Fig. 45. Vertical Section through "K-2" Triple Valve, Showing General Arrangement of Valves and Ports.

graduating valve. This transformation can be done at a minimum cost when the valves are returned to the works for heavy repairs.

A side view of the "K" triple valve, and the general arrangement of valves and ports, are shown in Figs. 45 and 46. Referring to Fig. 45, those parts which are different and not found in the standard triple, are as follows: *Valve Body* (1), *Slide-Valve* (2), *Graduating Valve* (3), *Retarding-Device Bracket* (4), *Retarding-Device Screw* (5), *Retarding-Device Washer* (6), *Retarding-Device Spring* (7), *Retarding-Device Stem Pin* (8), and *Graduating-Valve Spring* (9).

The *quick-service* feature gives a rapid serial operation of all brakes in service application. This is accomplished by using the principle of the standard triple in emergency applications—namely, discharging brake-pipe air into the brake-cylinder. That is, in service applications, some air from the brake-pipe passes into the brake-cylinder. The result is that the quick-service feature insures the operation of every brake, reduces the amount of air exhausted at the engineer's brake-valve and the possible loss of air due to flowing back through the feed-groove, and effects a saving of air.

The *retarded-release* feature operates so as to give practically a simultaneous release of all brakes in the train. This is accomplished by automatically restricting the exhaust of air from the brake-cylinder in the forward portion of the train, and allowing the others to release freely. This retarded release is due to the increased pressure which exists in the forward end of the brake-pipe when the brake-valve is in release position, and affects about the first thirty cars in the train.

The *uniform recharge* of the auxiliary reservoirs in the train is due to the fact that when the valve is in the retarded-release position, the ports connecting the brake-pipe with the auxiliary reservoir are automatically restricted. In other words, as long as the exhaust from the brake-cylinder is retarded, the recharge is restricted. This feature not only prevents the overcharging of the auxiliary reservoirs on the front cars, but, by drawing less air from the brake-pipe, permits the increase in brake-pipe pressure to travel more rapidly to the rear cars, where it is most needed for releasing and recharging those brakes.

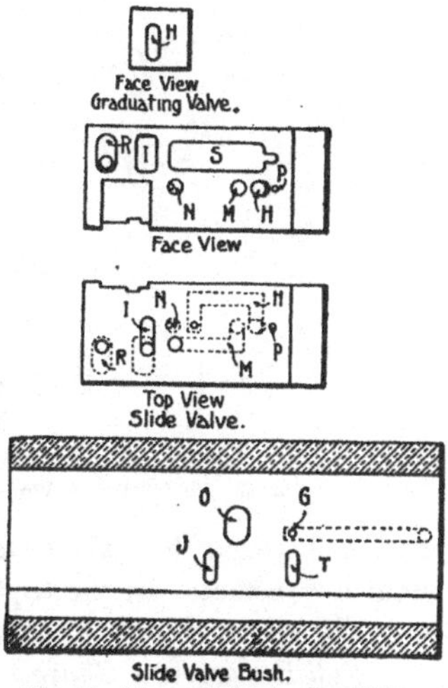

Fig. 46. Views of Graduating Valve, Slide-Valve, and Slide-Valve Bush of "K-2" Triple valve.

By reference to Fig. 46, which shows views of the graduating valve, slide-valve, and slide-valve bush, it will be seen that the ports are arranged along a longitudinal center line, making it very difficult to follow the course of air through them with a sectional view such as is shown in Fig. 45. For this reason, diagrammatic views shown in Figs. 47, 48, 49, 50, 51, and 52 are used in explaining the operation of the valve. In order to assist to a clearer understanding of the valve, the notation used to distinguish ports, valves, etc., is the same in all figures.

Referring to Fig. 45, the retarding-device brake (4) projects into the auxiliary reservoir; and its construction is such that free communication exists between the auxiliary reservoir and the chamber containing the slide-valve and the graduating valve. The graduating valve is of the slide-valve type, and moves over the top of the slide-valve, being carried along by the triple-valve piston. The

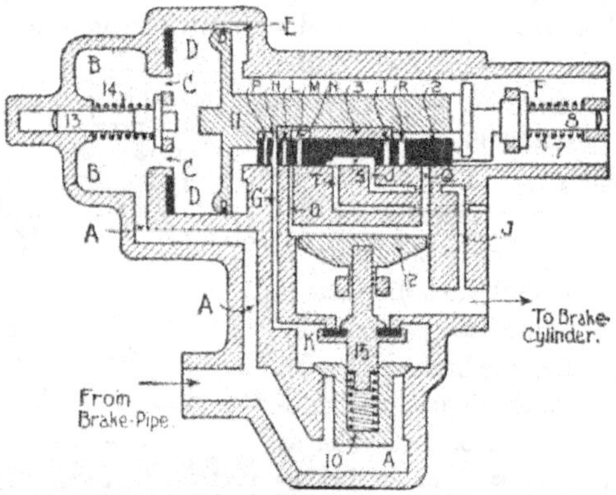

Fig. 47. "K" Triple Valve in Full-Release and Charging Position.

friction between the slide-valve and its seat prevents its movement until it is actuated by the triple-valve piston.

The operation of the "K" triple valve is as follows:

Full-Release and Charging Position. Fig. 47 shows the valve in this position. Air enters from the brake-pipe, and passes through the port *A* into the chamber *B*, through the ports *C*, into the cylinder *D*, through the feed-groove *E*, into the chamber *F* above the slide-valve, and finally passes into the auxiliary reservoir. The feed-groove *E* is the same size as that used in the standard triple. In the "K-2" triple, the port *H* is added to the slide-valve, through which air entering from the port *G* can feed into the auxiliary reservoir in order that a greater volume of air can be handled to supply the auxiliary reservoir of a 10-inch brake-cylinder. The port *H* is not placed in the "K-1" triple. Brake-pipe pressure, entering by the port *A*, lifts the check-valve (10), passes through the ports *G* and *H* into the chamber *F*, and thence into the auxiliary reservoir.

The process described above continues until the pressure in the auxiliary reservoir and brake-pipe become equal. The auxiliary reservoir is then said to be fully charged.

Quick-Service Position. In making a service application of the brakes, air is slowly exhausted from the brake-pipe, and the pressure in the chamber *D* is reduced. When the difference in the auxiliary reservoir and brake-pipe pressures is sufficient to overcome the friction

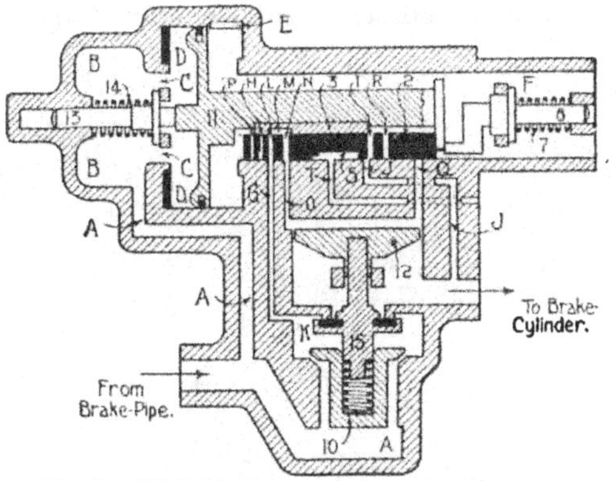

Fig. 48. "K" Triple Valve in Quick-Service Position.

of the piston (11) and the graduating valve (3), the piston moves to the left. As the piston (11) moves to the left, a shoulder on the right end of the piston strikes the right end of the slide-valve (2) and moves it to the left until the piston (11) strikes the end of the graduating stem (13). The parts of the valve then occupy the position shown in Fig. 48. In this position, air flows from the auxiliary reservoir into the chamber *F* through the ports *I* and *J*, into the brake-cylinder. At the same time, the small amount of air contained in the cavity *K* passes through the ports *G* and *L*, the cavity *M*, the ports *N* and *O*, around the emergency piston (12), into the brake-cylinder. When the pressure in the auxiliary reservoir drops below that in the brake-pipe, the check-valve (10) lifts, and air passes from the brake-pipe through the ports mentioned above, into the brake-cylinder. The emergency piston (12) fits loosely in its cylinder and permits air to pass around it without pressing it downward. The ports *G*, *L*, *N*, and *O* are proportioned so that there is no danger of any movement of the emergency piston (12). If this should occur, however, an emergency application would result.

It is readily seen that the action just described will greatly reduce the brake-pipe reduction necessary at the brake-valve, since air is taken into the brake-cylinder from the brake-pipe; also, that a higher cylinder pressure will result than if no air from the brake-pipe passed into the brake-cylinder.

Full-Service Position. In short trains, the volume of air in the brake-pipe is comparatively small. In service applications, air

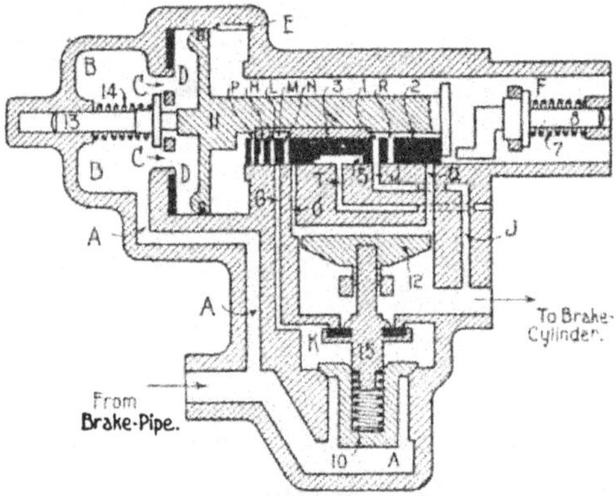

Fig. 49. "K" Triple Valve in Full-Service Position.

discharges so rapidly by the quick-service feature that an emergency would result were it not automatically prevented by the valve itself. In service applications, if the drop in brake-pipe pressure is more rapid than that in the auxiliary reservoir, then the valve takes the full-service position represented in Fig. 49. It will not, however, take the emergency position, because there is no sudden drop in the brake-pipe pressure. In the full-service position, the pressure behind the piston (11) is such that the graduating spring (14) is slightly compressed. This moves the slide-valve (2) to the left sufficiently to close the quick-service port *G*, and brings the port *I* into full registration with the port *J*. In this position, no air can enter into the brake-cylinder through the port *G*; but since the ports *I* and *J* are fully open, air is free to pass from the auxiliary reservoir into the brake-cylinder.

Lap Position. When the brake-pipe pressure has become constant after an application has been made, air continues to flow from the auxiliary reservoir through the ports *I* and *J* to the brake-cylinder, until the pressure in the chamber *F* becomes enough less than that in the chamber *D* to cause the piston (11) to move to the right. When the shoulder on the piston (11) strikes the left end of the slide-valve (2), it comes to rest on account of the frictional resistance of the slide-

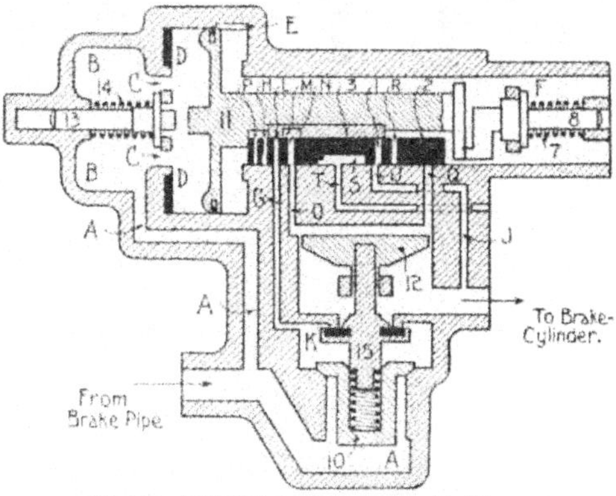

Fig. 50. "K" Triple Valve in Lap Position.

valve. In this position, all ports are closed and the valve is said to be *lapped* (see Fig. 50).

Retarded-Release and Charging Position. It is a well-known fact that in a freight train fitted with standard triples, the cars nearest the engine will release first when the engineer places the brake-valve in release position. This is due, *first*, to the friction of the air in the brake-pipe; and *second*, to the fact, that the auxiliary reservoirs of those brakes which release at the forward end begin to recharge, taking air from the brake-pipe, which reduces the pressure-head. The retarded-release feature overcomes the second point mentioned by taking advantage of the first. The friction of the air in the brake-pipe causes the pressure to build up more rapidly in the chamber *D* of triples at the front end of the train, than it does in those at the rear. When this pressure in the chamber *D* increases sufficiently above that in the auxiliary reservoir to overcome the frictional resistance of the piston, graduating valve, and slide-valve, all three parts move to the right until the piston strikes the retarding-device stem (8), which is held in position by the spring (7). The parts will then be in the position represented in Fig. 47. If, however, the pressure in the chamber *D* builds up faster than the auxiliary reservoir can recharge (as is the case if the triple is near the head of the train), then the piston moves still farther to the right, compressing the retarding-device spring (7) until the parts occupy the position shown in Fig. 51. In this position, the back

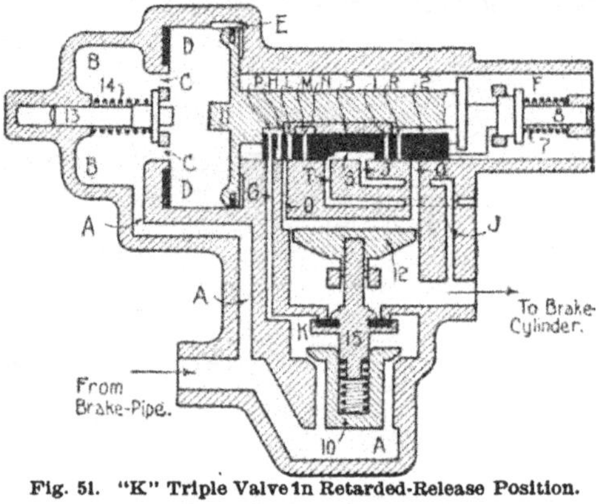

Fig. 51. "K" Triple Valve in Retarded-Release Position.

of the piston (11) is in contact with the slide-valve bush, and, acting as a valve, prevents any air from passing into the auxiliary reservoir through the feed-groove *E*; but the port *P* now registers with the port *G*, permitting air to pass from the chamber *A*—lifting the check-valve (10)—through the ports *G* and *P*, into the auxiliary reservoir. By this latter route, the auxiliary reservoir is recharged only about half as fast as it would be if charged through the feed-groove *E*. As the pressure increases in the auxiliary reservoir and becomes nearly equal to that in the chamber *D*, the retarding-device spring (7) overcomes the friction of the piston, slide-valve, and graduating valve, and moves them to the left to the position shown in Fig. 47. After this, recharging continues through the feed-groove *E* until the pressures are equalized. In the retarded-release position, the exhaust cavity S connects the port J with the exhaust port T, and the air in the brake-cylinder is discharged into the atmosphere. The discharge is very slow, however, since the small extension of the cavity S (see Fig. 46) is over the port T. This is the retarded-release feature, and affects about the first thirty cars in the train. Finally, when the valve takes the position shown in Fig. 47, the cavity S completely covers the port T, and a free discharge of air from the brake-cylinder occurs.

Emergency Position. This position is shown in Fig. 52. The operation of the "K" triple valve in emergency applications is the

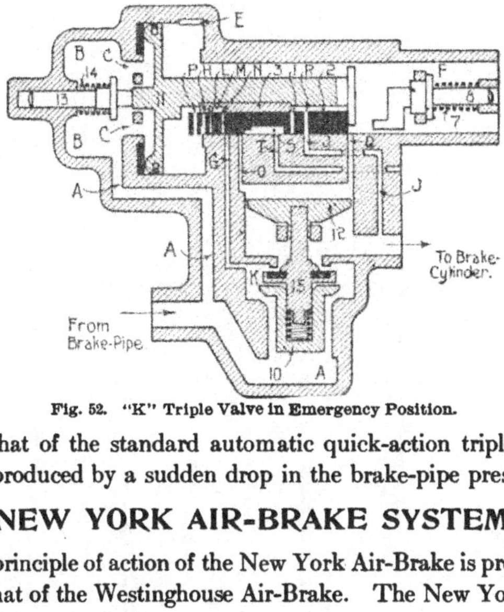

Fig. 52. "K" Triple Valve in Emergency Position.

same as that of the standard automatic quick-action triple. Quick action is produced by a sudden drop in the brake-pipe pressure.

NEW YORK AIR-BRAKE SYSTEM

The principle of action of the New York Air-Brake is precisely the same as that of the Westinghouse Air-Brake. The New York system is composed of the air-compressor, main reservoir, pump-governor, engineer's brake-valve, brake-pipe, triple valve, auxiliary reservoir, brake-cylinder, and pressure-retaining valve, which are the principal parts and are very similar to those used in the Westinghouse system. The only parts which need special explanation are the air-pump, engineer's brake-valve, and the triple valve.

New York Air-Pump. The New York Air-Pump is a duplex pump, and is built in two sizes. The larger size is shown in section in

Fig. 53. On the lower part are located the steam cylinders, each being 7 inches in diameter. The piston-rods connecting the steam pistons with the air pistons are made hollow for a portion of their length. This hollow portion provides a place for the stem which operates the steam valve. The action of the pump in compressing

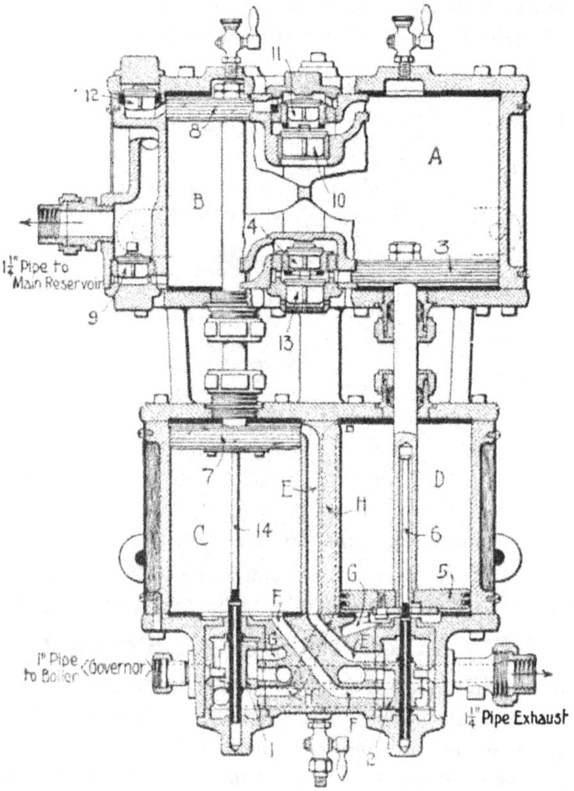

Fig. 53. Section of New York Air-Pump.

air is very similar to that of a compound steam engine, the air being compounded instead of steam. The entire valve-gear is very simple. The valves (1) and (2) controlling the action of the pistons are plain D slide-valves. The air-valves are simple check-valves. The operation of the pump in compressing air is as follows:

Each air cylinder fills with free air at every stroke. The pistons of one side rest while those on the other side are in motion. The valve on one side controls the supply of steam to the opposite side. In the position of the pistons shown in Fig. 53, the piston (3) has completed its stroke, and has forced the air in the cylinder *A* through the air-valve (4) into the cylinder *B* at about 40 pounds' pressure. The plate on the piston (5) has come in contact with the shoulder on the valve-stem (6), and moved the steam valve (2) to the position shown. This opens the port *E*, and steam is permitted to act on the top of the piston (7), forcing it downward. The steam below the piston (7) passes out through the port *F* into the exhaust pipe. As the piston (7) descends, the piston (8) is pulled downward, forcing the partially compressed air in the cylinder *B* out through the air-valve (9) into the main reservoir.

As the piston (8) descends, air at atmospheric pressure enters through the air-valves (10) and (11) and fills the space above the piston (8). In the same way, the cylinder *A* above the piston (3) is also filled with air entering through the air-valve (10). When the piston (7) reaches the lower end of the cylinder *C*, the valve stem (14) is moved downward and causes the steam valve (1) to uncover the port *G*. Steam is now permitted to act below the piston (5), causing it to rise and force the air above the piston (3) through the valve (11), into the cylinder *B* above the piston (8). As the piston (5) ascends, the steam in the cylinder *D* passes through the port *H* and the cavity in the valve (1), into the exhaust pipe. Air entering through the air-valve (13) fills the cylinder *A* below the piston (3). When the piston (5) reaches its highest point, the head on valve-stem (6) engages with the plate on the piston (5), and lifts the steam valve (2) until the port *F* is uncovered. The piston (7), now being at the bottom of its stroke, is acted on by steam from the port *F*, and is forced upward, discharging the air above the piston (8) through the air-valve (12) into the main reservoir. Air entering through the air-valves (13) and (4) fills the cylinder *B* below the piston (8). In this position, the plate on the piston (7) has lifted the valve-stem (14), causing the steam valve (1) to uncover the port *H*. Steam now acts on the top side of the piston (5) through the port *H*, forcing it downward and completing the cycle.

This type of air-pump is more efficient than the type represented by the nine and one-half inch Westinghouse air-pump, since the air cylinders are proportioned such that three measures of air are compressed for two measures of steam, whereas in the Westinghouse pump only two measures of air are compressed for two measures of steam.

New York Engineer's Brake Valve. The New York engineer's brake-valve performs the same functions as the standard Westinghouse engineer's brake-valve. It is illustrated in Figs. 54, 55, 56, and 57. Fig. 54 is a side view showing the different positions of the

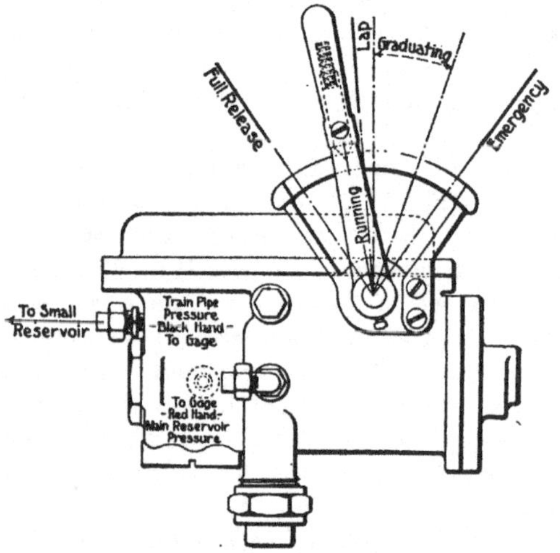

Fig. 54. Side View of New York Engineer's Brake-Valve, Showing Different Positions of Handle.

handle. Fig. 55 shows a longitudinal section of the valve, a plan of the valve-seat, and the face of the slide-valve. Fig. 56 is a section through the feed-valve as seen from the rear. Fig. 57 is a section through the slide-valve as seen from the front. The action of the valve when in its different positions is described as follows:

Running Position. Fig. 55 shows the position of the parts when the handle of the brake-valve is in running position. The main reservoir is in communication with the chamber A; and the brake-pipe, with the chamber B. The chamber C, to the right of the piston (1), is connected to a small reservoir. When the handle is in running position, the discharge ports *E*, *F*, and *G* in the slide-valve (2) are closed; and air from the main reservoir flows from the chamber *A*, lifting the feed-valve (3), passing through the port *H* (see Fig. 56), into the chamber *B*, and thence to the brake-pipe.

Service Position. In making service applications, the handle of the brake-valve is placed in one of the *service* or *graduating* notches

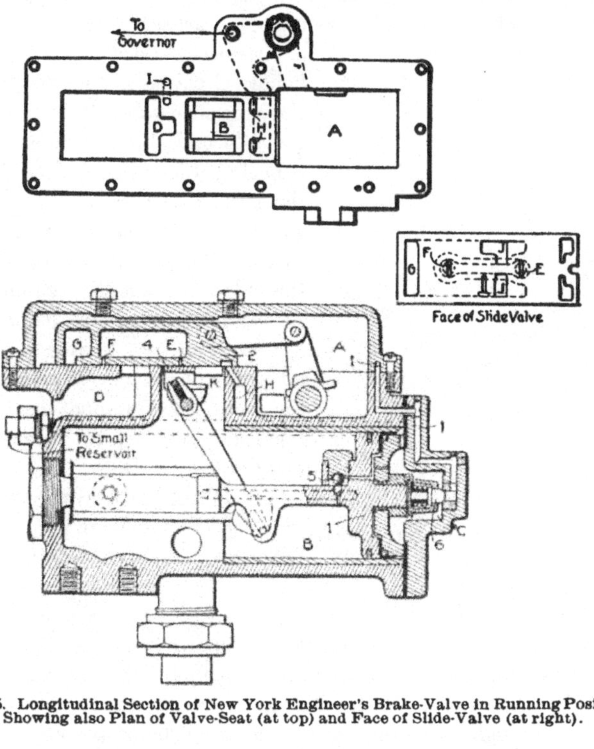

Fig. 55. Longitudinal Section of New York Engineer's Brake-Valve in Running Position, Showing also Plan of Valve-Seat (at top) and Face of Slide-Valve (at right).

illustrated in Fig. 54. Placing the handle in this position moves the slide-valve (2) to the right, uncovering the ports *E* and *F*, thus permitting brake-pipe air to escape from the chamber *B* to the atmosphere through the passage *D*. Air continues to be discharged into the atmosphere until the pressure in the brake-pipe and chamber *B* is decreased sufficiently to permit the pressure in the chamber *C* (which is in communication with the small reservoir) to move the piston (1) to the left. This movement operates the small slide-valve (4), moving it to the right and closing the port *E*. The small reservoir mentioned above receives its supply of air from the chamber *C*, which, in turn, is supplied with air from the chamber *B*, entering through the ball check-valve (5). For light applications, the first notches are used; and for heavier ones, th last notches. Full-service application is obtained when the handle is placed in the last service notch.

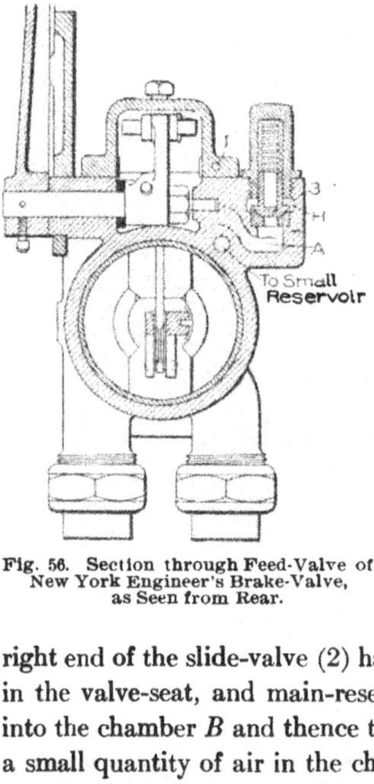

Fig. 56. Section through Feed-Valve of New York Engineer's Brake-Valve, as Seen from Rear.

Emergency Position. When the handle is placed in emergency position, the slide-valve (2) is moved to the right until direct communication is made between the chamber *B* and the exhaust passage *D*. In this position, air flows from the chamber *B* through the port *J* (see Fig. 57) in the slide-valve, out through the port *G*, and into the exhaust passage *D*.

Lap Position. In this position, all communication is closed between the main reservoir and the brake-pipe, and between the brake-pipe and the atmosphere.

Release Position. When the handle is placed in release position, the slide-valve (2) is moved to the extreme left. In this position, the right end of the slide-valve (2) has uncovered the port *K* (see Fig. 55) in the valve-seat, and main-reservoir air flows from the chamber *A* into the chamber *B* and thence to the brake-pipe. At the same time, a small quantity of air in the chamber *C* and the small reservoir discharges through the ports *I* and *J* into the exhaust passage *D*; and brake-pipe pressure, acting on the piston (1), moves it to the position shown in Fig. 55, ready for the next service application. The vent valve (6) controls the passage *I* leading to the valve-seat. The handle of the brake-valve should not remain in this position too long, as there is danger of the auxiliary reservoirs becoming overcharged.

If, after an application, the valve handle is placed in running position, the brakes will be released; but considerable time will be required, since the air must be supplied to the brake-pipe through the feed-valve (3).

New York Quick-Action Triple Valve. Fig. 58 shows the New York triple valve in section. Its action is quite similar to that of the Westinghouse triple valve. It differs in its quick-action feature, however, in that, when an emergency application is made, no addi-

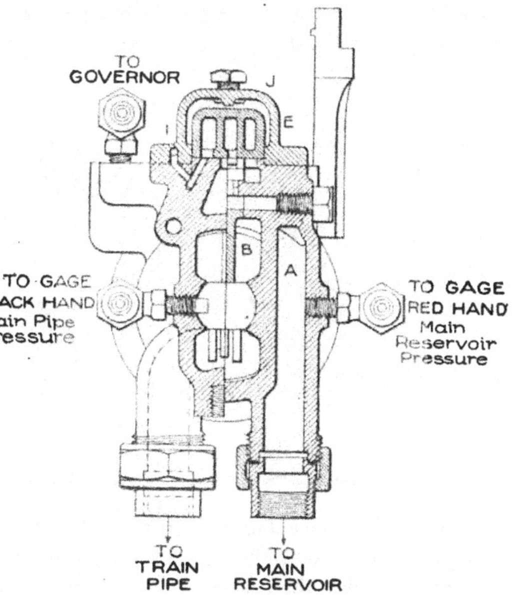

Fig. 57. Section through Slide-Valve of New York Engineer's Brake-Valve, as Seen from Front.

tional brake-cylinder pressure is obtained above that secured in a full-service application. The action of the valve in service and emergency application is as follows:

Charging and Release Position. The different parts of the valve are shown in this position in Fig. 58. Air from the brake-pipe enters the chamber A, passes through the ports B and C into the chamber D, through the feed-groove E into the chamber F, and into the auxiliary reservoir. Air continues to flow into the auxiliary reservoir until its pressure is the same as that in the brake-pipe. The head of the piston (1) is made so as to form a cylinder in which the piston (2) moves. Air at brake-pipe pressure enters the chamber *G* through the port *H*. If air-pressure exists in the brake-cylinder when the valve is in this position, it will flow out into the chamber *I* through the port *J*, the cavity *K*, and the port *L*, into the exhaust cavity *M*, to the atmosphere. In this position, air exhausts from the brake-cylinder until the brake is fully released.

Service Position. When the engineer's brake-valve is placed in service position, air is exhausted from the brake-pipe, and the pressure

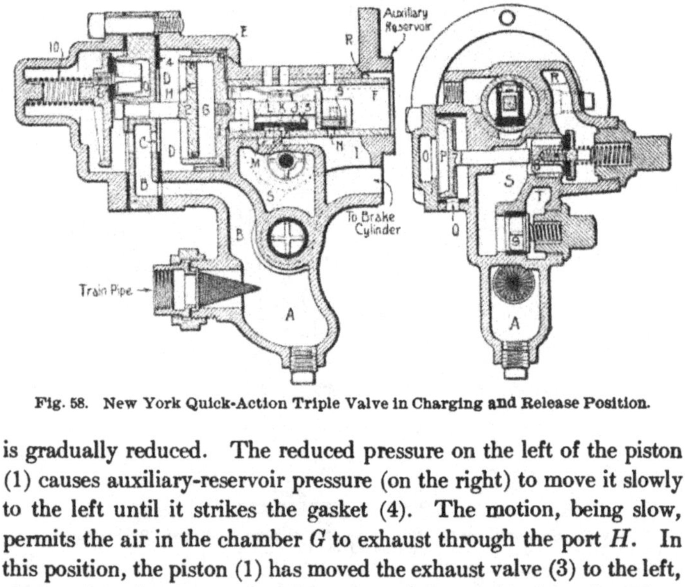

Fig. 58. New York Quick-Action Triple Valve in Charging and Release Position.

is gradually reduced. The reduced pressure on the left of the piston (1) causes auxiliary-reservoir pressure (on the right) to move it slowly to the left until it strikes the gasket (4). The motion, being slow, permits the air in the chamber *G* to exhaust through the port *H*. In this position, the piston (1) has moved the exhaust valve (3) to the left, closing the exhaust port *J*, and has caused the graduating valve (5) to uncover the port *N*. Air now flows from the auxiliary reservoir, through the port *N*, to the chamber *I*, into the brake-cylinder.

Lap Position. If the brake-pipe reduction has not been sufficient to cause full equalization of the auxiliary-reservoir and brake-cylinder pressure, air will continue to flow from the auxiliary reservoir to the brake-cylinder until the pressure on the left of the piston (1) moves it toward the right. This movement of the piston (1) is stopped when the left shoulder on the piston (1) strikes the left end of the exhaust valve (3). In this position, the port J is closed by the slide-valve (3), port N is closed by the graduating valve (5), and the valve is said to be *lapped.*

Emergency Position. The piston (1) has the same movement in both service and emergency positions. The port H is of such size that when the piston (1) moves slowly to the left, as in service applications, the air in the chamber G is forced out without moving the piston (2) from the position shown. If an emergency application is desired, the handle of the engineer's brake-valve is moved at once to emergency position. This causes the brake-pipe pressure to drop very suddenly, and the piston (1) to move to the left so rapidly that the air in the chamber G cannot discharge through the port H fast enough to prevent the piston (2) from being disturbed. The result is that the piston (2) is moved to the left. This movement causes the valve (6) to be momentarily pushed from its seat by the stem of the piston (2). This allows air from the brake-pipe to enter the cavity O, flow around the side to the chamber P, and escape to the atmosphere through the port Q. The air now in the chamber P forces the piston (7) to the right, which unseats the valve (8), and permits air from the auxiliary reservoir to flow through the port R, the valve (8), the chamber S, the check-valve (9), and the chamber T, into the chamber I, and thence to the brake-cylinder. As the last-mentioned passages are very large, full braking pressure is obtained instantaneously. While the action just described is going on, air from the chamber G is being discharged through the port H. When it is entirely exhausted, the spring (10) seats the valve (6), and all parts occupy positions as described under *service position.*

FOUNDATION BRAKE-GEAR

The foundation brake-gear includes all levers, rods, beams, pins, etc., which serve to transmit the braking force from the piston of the brake-cylinder to the brake-shoes. It is important that all longitudinal rods should be parallel with the center line of the car when the brakes are fully applied. The brake-beams should be hung in such a manner that they will always be the same distance above the rail, the reason being that this practice reduces the chance for flat wheels, since the piston travel is not affected by the loading or unloading of the car.

The rods and levers should be designed so that they will move in the same direction when the brakes are applied by hand as when by air. The levers should stand approximately at right angles to the rods when the brakes are set.

A number of different systems of rods and levers have been used by different railroad companies, with varying degrees of success. The systems adopted by the Master Car-Builders' Association are diagrammatically shown in Fig. 59. The four cases shown represent two general systems—those where the brake-shoes are *hung inside*, between the truck wheels, and those where they are *hung outside*. Freight-

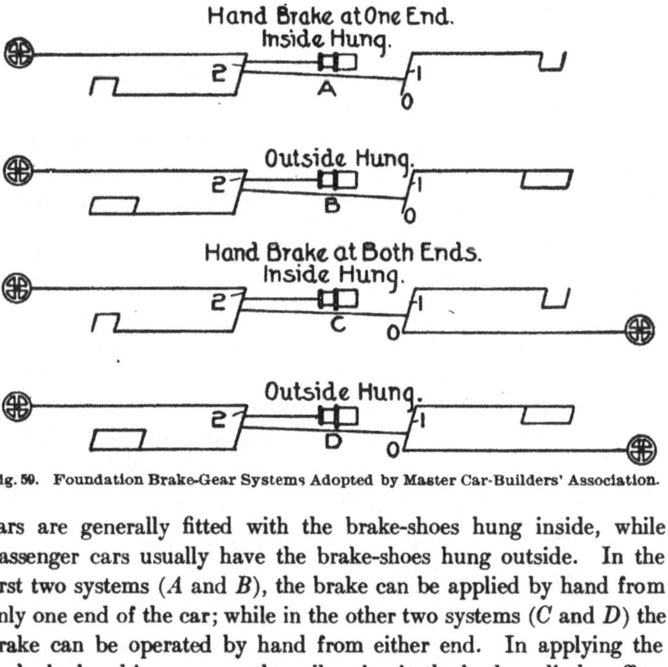

Fig. 59. Foundation Brake-Gear Systems Adopted by Master Car-Builders' Association.

cars are generally fitted with the brake-shoes hung inside, while passenger cars usually have the brake-shoes hung outside. In the first two systems (*A* and *B*), the brake can be applied by hand from only one end of the car; while in the other two systems (*C* and *D*) the brake can be operated by hand from either end. In applying the brake by hand in any case, the coil spring in the brake-cylinder offers no resistance, since the push rod has no pin connection to the piston rod. The piston-rod of the brake-cylinder is hollow. When the brake is operated by hand, the push rod slides outward in the hollow rod without moving the piston. A detailed description of the operation of the four systems shown is not thought necessary. One or two points, however, might assist to a clearer understanding of them. The lower end of the lever (1) in the systems *A* and *B* is fixed at *O*. The lower end of the lever (1) in the systems *C* and *D* is held by a stop at *O*, and cannot move to the left, but is free to move to the right, when the brake is operated by hand from the right-hand end of the car. The lever (2) in all four systems has no fixed points. In all cases, the arrangement is such that no brake-shoe will press against its wheel with any great force until all brake-shoes are held firmly against their respective wheels, and all shoes press against the wheels with an equal force.

Fig. 60, with all parts named, shows the application of the system *A* to a freight-car. No explanation is needed.

Leverage. It is a well-known principle in Mechanics, that the greater the weight on a car wheel, the greater the brake-shoe pressure on that wheel necessary to cause it to slide on the track. For this reason, in designing the brake-rigging for a car, the light, or unloaded, weight of the car is the basis of all calculations. If the loaded weight of the car were used in calculating the levers, the proportions would be such that if the brakes were applied

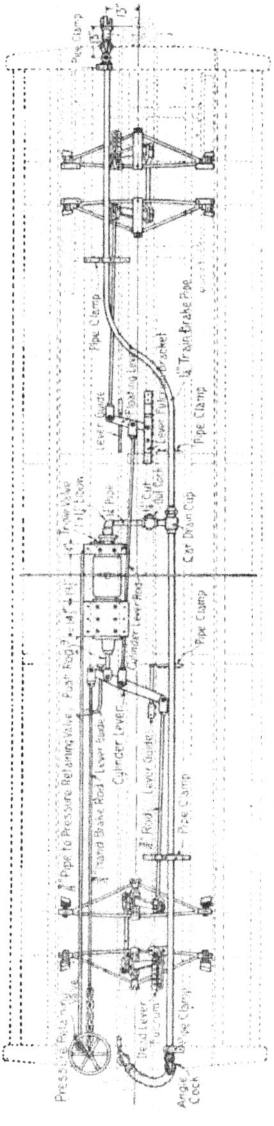

Fig. 60. Application, to a Freight Car, of Inside-Hung Brake-Gear System Shown at *A* in Fig. 59.

A

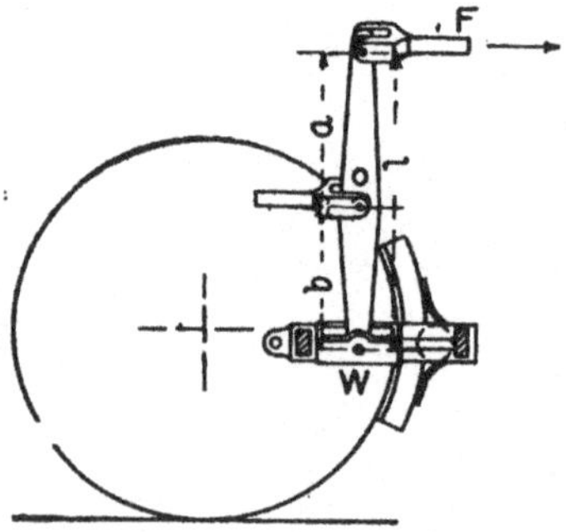

$$W = \frac{F \times a}{b};$$

$$F = \frac{W \times b}{a}; \quad l = a + b;$$

$$a = \frac{W \times b}{F}; \text{ or, } a = \frac{W \times l}{F + W};$$

$$b = \frac{F \times a}{W}; \text{ or, } b = \frac{F \times l}{F + W}.$$

FULCRUM BETWEEN APPLIED AND DELIVERED FORCES.

B

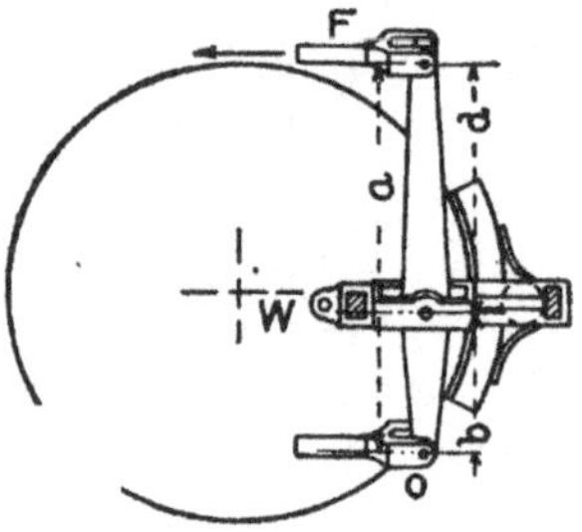

$$W = \frac{F \times a}{b};$$

$$F = \frac{W \times b}{a}; \quad a = b + d;$$

$$a = \frac{W \times b}{F}; \text{ or, } a = \frac{W \times d}{W - F};$$

$$b = \frac{F \times a}{W}; \text{ or, } b = \frac{F \times d}{W - F}.$$

DELIVERED FORCE BETWEEN FULCRUM AND APPLIED FORCE.

C

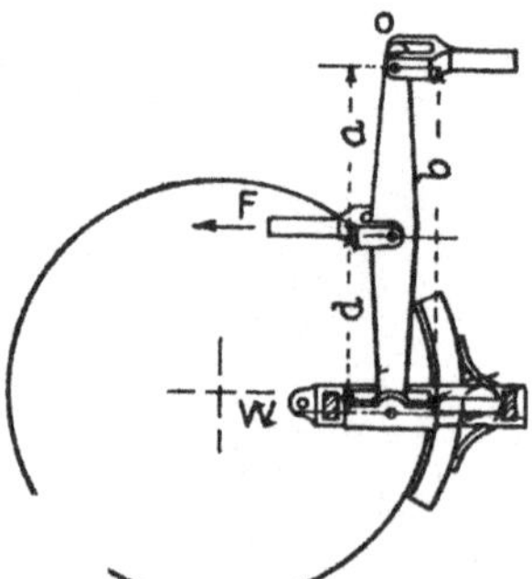

$$W = \frac{F \times a}{b};$$

$$F = \frac{W \times b}{a}; \quad b = a + d;$$

$$a = \frac{W \times b}{F}; \text{ or, } a = \frac{W - d}{F - W};$$

$$b = \frac{F \times a}{W}; \text{ or, } b = \frac{F \times d}{F - W}.$$

APPLIED FORCE BETWEEN FULCRUM AND DELIVERED FORCE.

Fig. 61. Illustrating Application of Principle of Moments to Levers in Brake Systems.

when the car was unloaded, the wheels would slide. In order to prevent any chance arising of having flat spots worn on the wheels, due to the wheels sliding on the track, the following percentages of light weights on the wheels are usually employed in determining the brake-shoe pressure:

Passenger cars.	90 per cent.
Freight cars.	70 per cent.
Tenders.	100 per cent.
Locomotive drivers..	75 per cent (of weight upon the drivers).
Locomotive truck.	75 per cent (of weight upon the truck).

These percentages are sometimes changed to meet special conditions which arise.

In calculating the brake-shoe pressure of any car, one must know three things: *First*, the diameter of the brake-cylinder and its maximum pressure; *second*, the sizes and positions of all levers in the system; and *third*, a knowledge of the theorem of moments as used in Mechanics.

The principle or theorem of moments may be stated thus: *The product of the force applied at one pin and its perpendicular distance from the fulcrum pin, is equal to the product of the force delivered at the other pin and its perpendicular distance from the fulcrum pin.* This principle has been applied to the three different classes of levers; and the forces and distances have been worked out, and are shown in Fig. 61. The chief difficulty the beginner encounters is in locating the fulcrum pin. In A, B, and C (Fig. 61), the fulcrum pin is located at O, the force applied is F, and the force delivered is W. In any case, if the pull F on the lever is known, the brake-shoe pressure W can be determined.

Fig. 62 represents diagrammatically the scheme of levers and rods commonly used on freight-cars. All distances of rods from the center line of the car are taken when the levers are at right angles to it. The brake-cylinder is 8 inches in diameter, and has an area of about 50 square inches. If the maximum brake-cylinder pressure in emergency applications is 60 pounds, the total pressure delivered to the push rod would be $50 \times 60 = 3{,}000$ pounds. This 3,000 pounds is transmitted to the lever E at the pin (1). The lever E is of the class shown in B (Fig. 61), and its fulcrum is at the pin (3). Applying the formula gives 4,500 pounds delivered at the pin (2). This 4,500 pounds is

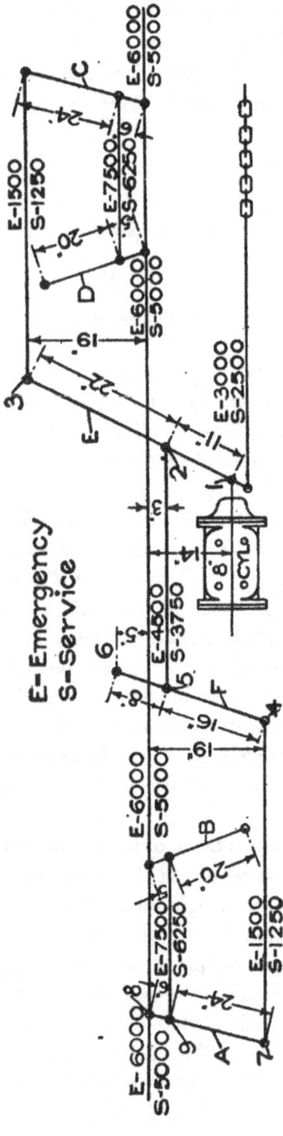

Fig. 62. Scheme of Levers and Rods Commonly Used on Freight-Cars.

transmitted to the lever *F*, which is of the class shown in *C* (Fig. 61), and its fulcrum is at the pin (6). Applying the formula gives 1,500 pounds delivered at the pin (4). This 1,500 pounds is transmitted to the lever *A*, which is of the class shown in *A* (Fig. 61), and its fulcrum is at the pin (9). Applying the formula gives 6,000 pounds delivered to the brake-beam at the pin (8). In a similar manner the other brake-beam pressures can be determined. In the figure, the calculation has been carried through for both service and emergency applications.

It is seen that 6,000 pounds is transmitted to the middle of each of the four brake-beams. Each brake-shoe will then receive a pressure of 3,000 pounds. Since there are eight wheels, the total braking pressure will be $8 \times 3,000 = 24,000$ pounds. This total braking pressure must not exceed 70 per cent of the unloaded weight of the car.

Automatic Slack-Adjuster. Full braking pressure will be secured as long as the maximum allowable brake-cylinder pressure can be maintained. Since the brake-cylinder pressure depends upon the length of stroke of the piston, it follows that the stroke of the piston should be kept as nearly constant as possible. The greater the stroke, the less the pressure. The stroke of the piston should be kept at about 8 inches. As the brake-shoes and various connections wear, the stroke of the piston is increased, and the pressure with which the shoes are forced against the wheels is decreased. In order to compensate for this wear, some means must be provided for taking up the

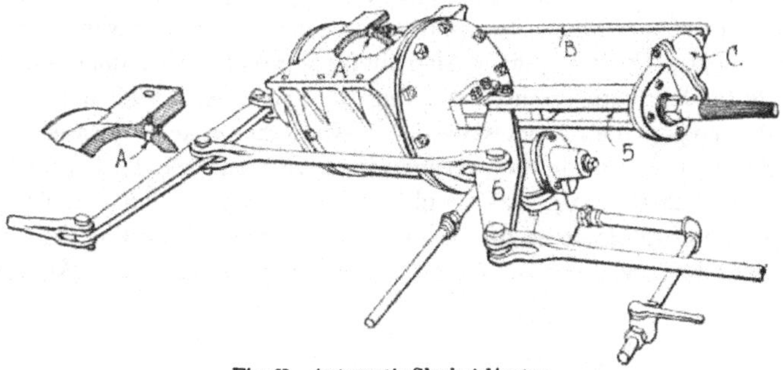

Fig. 63. Automatic Slack-Adjuster.

slack. This is done in one of two ways, either by changing the fulcrum pin of the dead lever (see Fig. 60) or by using the *automatic slack-adjuster*. The first method of adjustment is the one most commonly used, and is necessarily very coarsely graded. The *automatic slack-*

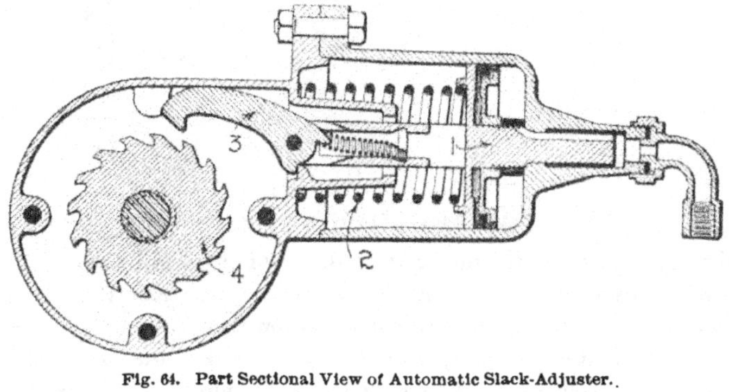

Fig. 64. Part Sectional View of Automatic Slack-Adjuster.

adjuster, when used at all, is usually fitted to the passenger-car equipment.

The *automatic slack-adjuster*, illustrated in Figs. 63 and 64, is manufactured by the Westinghouse Air-Brake Company. The purpose of the apparatus is to maintain a constant, predetermined piston travel. The brake-cylinder piston acts as a valve to control the admission and release of air to the pipe *B* through the port *A*. Whenever the stroke of the brake-cylinder piston is so great that the port *A* is passed by the piston, air from the cylinder enters the port *A* into the pipe *B*, and enters the cylinder *C*, which is shown in section in Fig. 64. The air entering the small cylinder acts on the piston (1) forcing it to the left, compressing the spring (2), and causing the small pawl (3) to engage the ratchet wheel (4). When the brake is released, the brake-cylinder piston returns, and air in the small cylinder *C* escapes to the atmosphere through the pipe *B* and the port *A*, thus permitting the spring (2) to force the piston (1) to its normal position. In so doing, the pawl (3) turns the ratchet wheel (4) on the screw (5), and

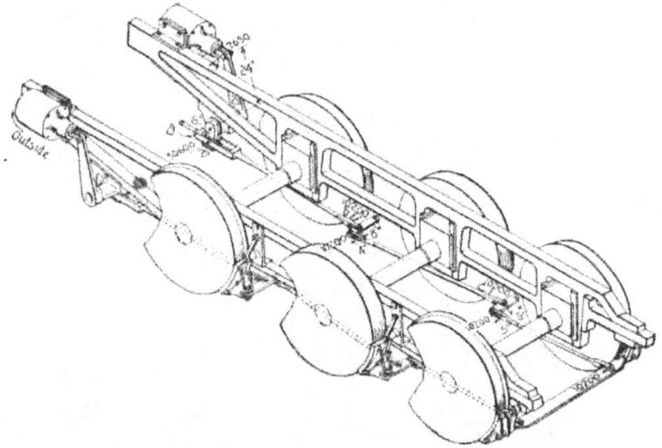

Fig. 65. Outside Equalized Driver-Brake for Locomotives.

thereby draws the fulcrum end of the lever (6) slightly nearer the slack-adjuster cylinder *C*. Each operation of the piston (1), as just described, reduces the brake-cylinder piston travel about $\frac{1}{32}$ of an inch. When the piston (1) is in its normal position, the outer end of the pawl (3) is lifted, permitting the screw (5) to be turned by hand.

Locomotive-Driver Brakes. The brakes are applied to the drivers of a locomotive in two general ways—by the *outside equalized* system, as illustrated in Fig. 65; and by *cams*, as shown in Fig. 66. The former scheme has practically replaced the latter because of its

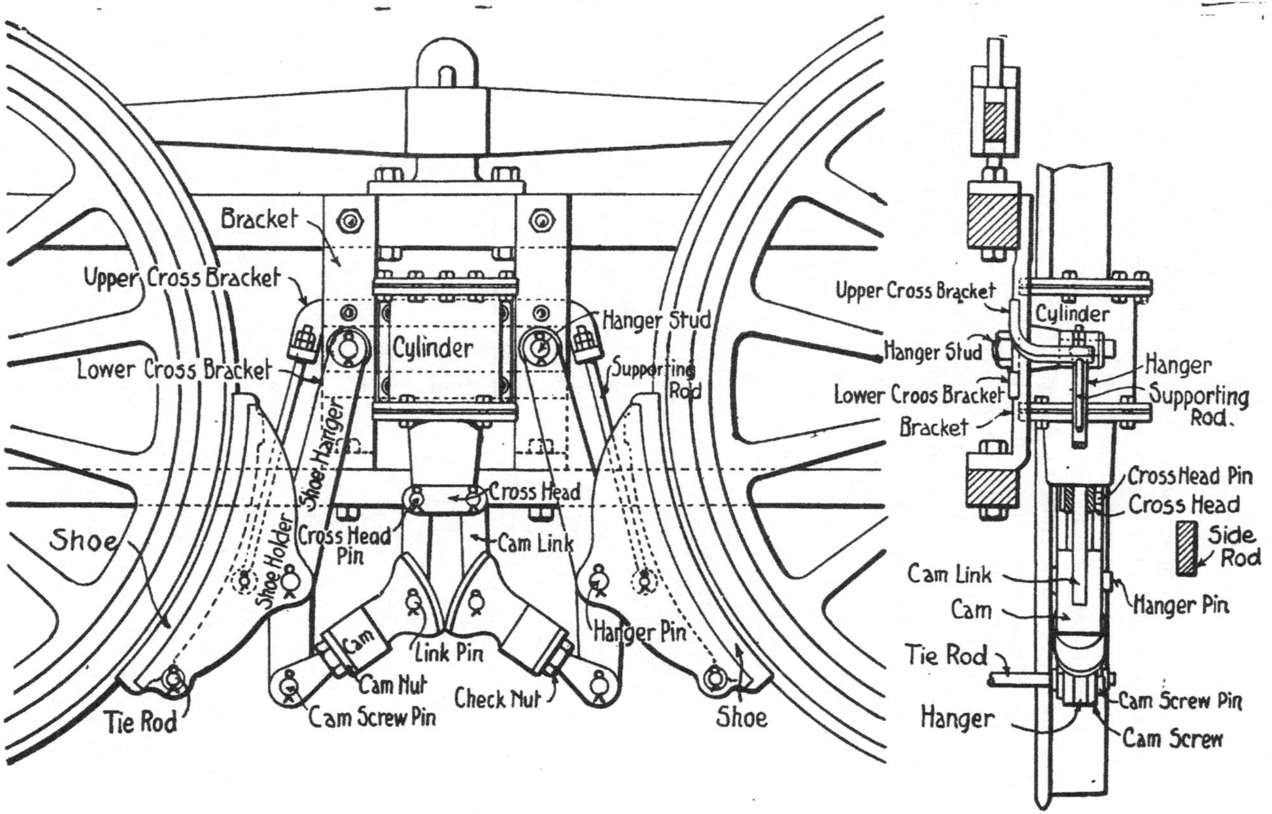

Fig. 66. Cam Driver-Brake for Locomotives.

being simpler in design and adjustment. The brake-cylinders and auxiliary reservoirs used on locomotives are usually proportioned so that the pressure in the brake-cylinder will equalize at 50 pounds. In the system shown in Fig. 65, the levers are constructed so that each wheel receives the same braking pressure. If the brake-cylinder is 14 inches in diameter and the cylinder pressure is 50 pounds, the pressure delivered at the pin *A* is about 7,650 pounds, while that on each wheel is 10,200 pounds. These values, of course, are different for different locomotives. The stroke of the piston is regulated by the adjusting mechanism at *B*.

The action of the cam driver-brake is shown in Fig. 66. When air is admitted to the brake-cylinder, the piston is forced downward.

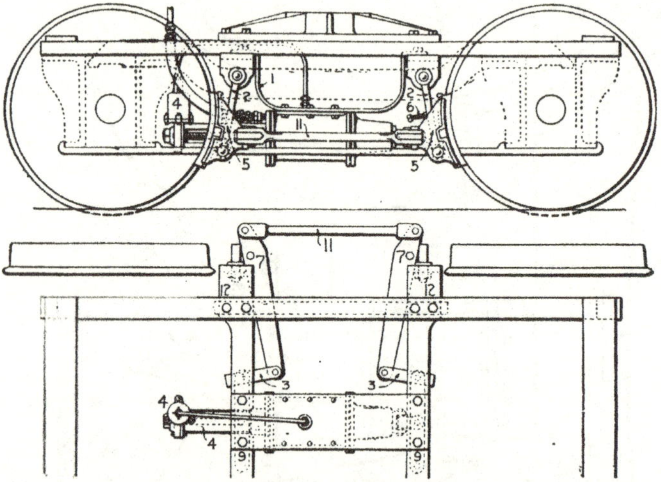

Fig. 67. Locomotive-Truck Brake.

This action pushes down the crosshead cams, which force the brake-shoes against the drivers. The piston travel is controlled by adjusting the cam nut on each cam.

Locomotive-Truck Brake. In certain types of locomotives, a considerable proportion of the weight of the locomotive is carried on the truck. It follows, that in order to develop the full braking power of the locomotive, a well-designed truck brake should be provided. The type of brake shown in Fig. 67 is frequently used. It is fitted with an automatic slack-adjuster. This feature is not so important here as on the car equipment.

WESTINGHOUSE TRAIN AIR-SIGNAL SYSTEM

The train signal system is very essential in maintaining fast schedules with passenger trains. Its object is to furnish a means of communication between the trainmen and enginemen. It is made up of the following principal parts:

1. A ¾-inch *signal pipe,* which extends throughout the length of the train, being connected between cars by flexible hose and suitable couplings.
2. A *reducing valve,* which is located on the engine, and which feeds air from the main reservoir into the signal pipe at 40 pounds' pressure.
3. A *signal valve and whistle,* located in the cab and connected to the signal pipe.
4. A *car discharge valve,* located on each car, which is connected to the signal pipe.

The action of the signal system is automatic. If an accident happens to the train, which breaks the signal pipe, the pressure in the signal pipe is reduced, and the whistle in the cab blows a blast. The trainmen signal the enginemen by opening the card ischarge valve, which reduces the pressure in the signal pipe. This reduction of pressure in the signal pipe operates the signal valve in the cab, which admits air to the whistle. The operation of the various parts is as follows:

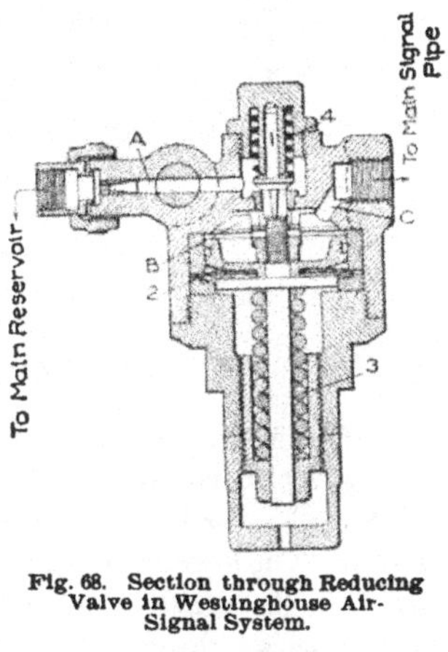

Fig. 68. Section through Reducing Valve in Westinghouse Air-Signal System.

Reducing Valve. A section through the reducing valve is shown in Fig. 68. This valve is located in a suitable place on the locomotive. Its purpose is to receive air from the main reservoir and feed it into the signal pipe, maintaining a pressure of 40 pounds. When no air is in the system, the parts occupy the position shown. When air is admitted from the main reservoir, it flows through the passage *A* and the supply valve (1), into the chamber *B* and out through the port *C* into the main signal pipe. When the air in the main signal pipe attains a pressure of 40 pounds, the pressure in the chamber *B*, acting on the piston (2), forces it downward, compressing the spring (3). This permits the spring (4) to close the supply valve (1). No more air can then enter the signal pipe until its pressure becomes reduced so that the spring (3) will force the piston (2) upward and lift the supply valve (1).

Signal Valve. The signal valve controls the supply of air to the whistle. Whenever a reduction of air-pressure occurs in the signal pipe, the signal valve admits air to the whistle. A section of the valve is shown in Fig. 69. The two compartments *A* and *B* are divided by

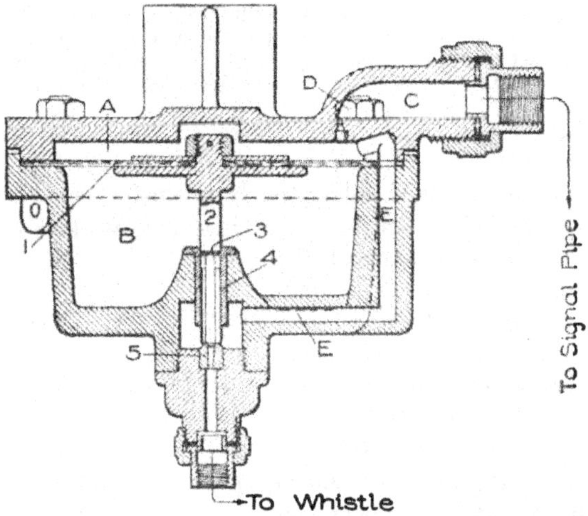

Fig. 69. Section through Signal Valve in Westinghouse Air-Signal System.

the diaphragm (1) to which is attached the stem (2). The stem (2) is milled triangular in section from the lower end to the peripheral groove (3). Above the groove (3), the stem (2) fits the bush (4) snugly. The lower end of the stem (2) acts as a valve on the seat (5). Air enters the signal valve from the signal pipe, through the passage *C*. It then passes through the small port *D* into the chamber *A*, and through the passage *E*, around the stem (2), into the chamber *B*. This charges the chambers *A* and *B* to signal-pipe pressure. A sudden reduction in signal-pipe pressure reduces the pressure in the chamber *A*; and the diaphragm (1), acted on by the pressure in the chamber *B*, rises, lifting the stem (2) and momentarily permitting air to pass from the signal pipe to the whistle. The resulting blast of the whistle is a signal to the enginemen. This same reduction of pressure in the signal pipe causes the reducing valve to recharge the system. The pressure in the chambers A and B equalizes quickly, and the lower end of the stem (2) returns to its seat.

Car Discharge Valve. The discharge valve is usually located outside the car, above the door, in such a position that the signal cord passing through the car can easily be fastened to the small lever of the valve.

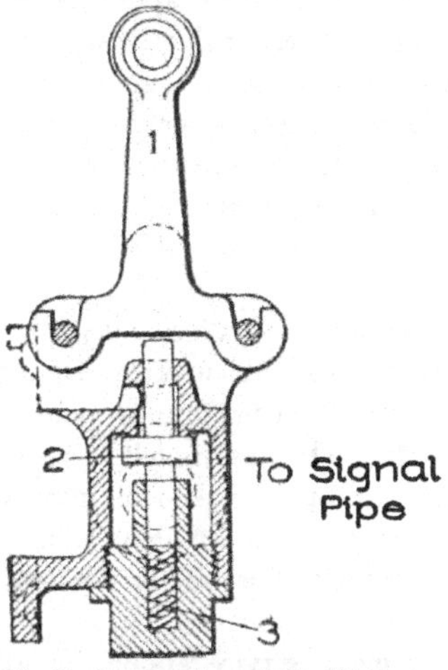

Fig. 70. Section through Car Discharge Valve in Westinghouse Air-Signal System.

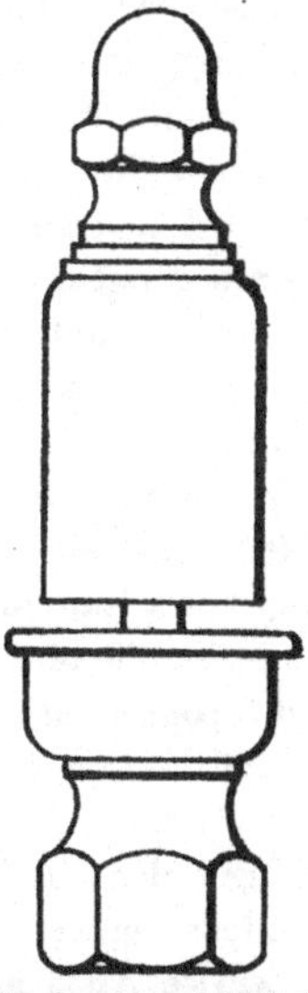

Fig. 71. Signal Whistle in Westinghouse Air-Signal System.

Fig. 70 is a section of the valve. The valve is connected to a branch pipe which extends from the signal pipe. The signal cord is connected to the eye in lever (1). Each pull in the signal cord causes the lever (1) to open the check-valve (2), permitting air to escape from the signal pipe. This causes a reduction in the signal pipe, which, in turn, causes the whistle to blow as previously described. The spring (3) closes the valve (2) when the signal cord is not held.

For the successful operation of the signal system, the signal pipe must be perfectly tight. Care must be exercised in using the car discharge valve, that sufficient time is permitted to elapse between successive discharges.

SPECIAL INSTRUCTIONS IN USE AND CARE OF AIR-BRAKE EQUIPMENT

Train Inspection. When a train is made up at a terminal, the air hose should all be coupled, and the angle-cocks all opened except the one at the rear end of the last car. The brake-pipe should then be charged to about 40 pounds, in order that the inspector may examine for leaks. When the brake-pipe has been fully charged, the engineer should apply the brake by making a light reduction in the brake-pipe, which should then be followed by a full-service application. He should note the time required in making these reductions, in order to be assured that all pistons are moved past the leakage groove when the train is out upon the road. The engineer, after making the full reduction, should leave his brake-valve in lap position until the inspector has examined the brake under every car. It should be the duty of the engineer to see that the brake equipment on the locomotive is in proper working order.

Running Test. In passenger service, when a locomotive has been changed or a train made up, the engineer should make a running test within one mile of the station, as follows: A brake-pipe reduction of about 5 pounds should be made. If the brakes are felt to be applying, and the time of the discharge is proportional to the number of cars in the train, the engineer will conclude that the brake is in proper working order. It is well, also, to make this test on approaching hazardous places.

Service Applications. In making a service application of the brakes, the first reduction should be about 5 pounds on a train of 30 cars or less, and about 7 pounds on a train exceeding 30 cars. This will insure the travel of all pistons beyond the leakage groove. Subsequent reductions of from two to three pounds can be made, to increase the braking power, if desired. A reduction of 25 to 30 pounds will make a full-service application. This should seldom be made, as it requires some time to recharge the system and release the brakes.

In stopping a passenger train, two applications should be used: the first should reduce the speed of the train to about 8 miles an hour when the train is within two or three car lengths of the point at which the train is to be stopped. Moving the brake-valve handle to release position for only sufficient time to release all brakes, then returning it to lap position, will make it possible for a second light application to stop the train. Just before all stops of passenger trains, except exact-position stops at water stations and coal chutes, the brakes should be released to avoid shocks to passengers. This release should be made on the last revolution of the drivers. If it should be made too soon, and the train keep on moving, the engineer's brake-valve should be moved to service position until the train stops.

In making stops of freight trains, the best practice is to shut off the steam, and allow the slack to run in before applying the brakes. The stop should be made with one application of the brakes. After the first reduction is made, if there are any leaks in the brake-pipe, the braking force will be increased, and any subsequent reduction should be made less, in order to make up for these leaks.

In stopping a long freight train at water stations and coal chutes, it is best to stop short of the place, cut off, and run up with the locomotive alone.

On a freight train, where the locomotive is not equipped with the straight air-brake, the brakes should not be released when the speed of the train is 10 miles per hour or less. If this is done, the brakes in the front of the train will release, and, as the slack runs out, the train may part. If the locomotive is equipped with straight air, the train brakes can be released after the locomotive brakes are set, without danger of parting the train. This can also be accomplished by the use of the Westinghouse "E T" equipment.

Emergency Applications. The emergency application should never be used, except in case of an emergency. If the necessity arises, an emergency application may be made after a service reduction of about 15 pounds.

In case an emergency is caused by the train parting, hose bursting, or the conductor's valve being opened, the engineer should place his valve on *lap*, in order to save the main-reservoir air.

Use of Sand. The use of sand increases the braking power of a train, and should be made in emergency stops. If sand is used in service stops, it should be applied some time before the brakes are applied, in order to have sand under the entire train. If, for any reason, the wheels should skid, do not apply the sand, as it will produce flat spots on the wheels.

Pressure-Retaining Valve. In holding trains on grades, a part or all of the retaining valves are set to hold 15 pounds in the brake-cylinder. If only part are set, those in the front of the train should be used.

Backing Up Trains. In backing up freight trains, the train should be stopped by the hand-brakes on the leading end of the train, for the reason that if air were used, the brakes would apply on the cars near the engine and the leading cars might cause a break-in-two.

In backing up a passenger train, where the train is controlled by a man on the leading car by means of an angle-cock, the engineer's valve should be in running position. This gives the man on the rear of the train full control of the brakes. As soon as the engineer feels the brake apply, he should place his valve on *lap*.

Double-Heading. When two or more locomotives are coupled in the same train, the brakes are operated by the leading locomotive. The cut-out cocks in the brake-pipe just below the engineer's valve on all locomotives but the first, should be closed. The pumps on all engines should be kept running.

Conductor's Brake-Valve. A conductor's brake-valve is located on each passenger car. The purpose of this valve is that the conductor may stop the train in case of emergency; if the engineer's brake-valve should fail to operate, he may signal the conductor to apply the brakes by opening the valve.

Use of Angle-Cocks. In setting a car out of a train, first release the brakes, then close the angle-cock on both sides of the hose to be disconnected, and finally disconnect the hose by hand. Before leaving a car on the side track, the air-brakes should first be released by opening the release valve on the auxiliary reservoir; and if the car is on a grade, the hand-brake should be set.

The angle-cock should not be opened on the head end of a train while the locomotive is detached. When connecting a locomotive to a train that is already charged with air, the angle-cock at the rear of the tender should be opened first, to allow the hose to become charged and thus prevent a slight reduction in the brake-pipe, which might set the brakes. All angle-cocks upon charged brake-pipes should be opened slowly.

Cutting Out Brakes. If the brake equipment on any car is defective, it may be cut out by closing the cut-out cock in the branch pipe leading from the brake-pipe to the triple valve. The release valve on the auxiliary reservoir should be opened to discharge the air. Never more than three cars with their brakes cut out should be placed together in a train, on account of the emergency feature being unable to skip more than this number.

Air-Pump. The air-pump should be run slowly with the drain-cocks open until the steam cylinder becomes warm and sufficient air-pressure has been attained to cushion the air, after which time the throttle may be fully opened. The lubricator should be in operation as soon as possible after starting, and the swab on the piston-rod should be kept well oiled. The air cylinder should receive oil each trip. Valve oil should be used, and it should be inserted through the oil-cup provided for that purpose, and not through the air strainer.

Engineer's Brake-Valve. With the handle in running position, the main-reservoir pressure should be maintained at 90 pounds, and the brake-pipe at 70 pounds. This requires that the springs in the pump-governor and feed-valve must be carefully adjusted, and that no leaks exist between ports in the rotary valve. The rotary valve should be cleaned and oiled when necessary; and if leaks exist, the valve should be scraped to a fit.

Triple Valve and Brake-Cylinders. These should receive an occasional cleaning and oiling, in order that they may be relied upon to fulfil their function. In cleaning the cylinder, special attention should be given to removing any deposit in the leakage groove. The walls of the cylinder should be coated with suitable oil or grease, and all bolts in the cylinder-head and follower should be kept tight.

In cleaning the triple valve, a common practice is to place the removable parts in kerosene until the other parts and the brake-cylinder have been cleaned. The parts are then removed, cleaned, oiled, and replaced. Special care should be given to the slide-valve and its seat, and to the graduating valve. All lint should be removed before replacing the parts. The piston packing-ring should never be removed, except for renewing. A few drops of oil is all that is necessary for lubricating the entire triple valve. No oil should be permitted to get upon the gaskets or rubber-seated valve. The graduating-valve and check-valve springs should be examined, and, if necessary, renewed.

AIR-BRAKES AS APPLIED TO ELECTRIC CARS

That electric street-cars and interurban cars should be equipped with reliable and efficient braking apparatus, is a well-established fact. It is emphasized by the frequent accidents which occur on roads where poorly constructed braking appliances are used. The modern electric car is several times heavier than cars used a decade ago, and speeds have increased remarkably, yet we frequently find cars fitted with braking apparatus but little better than that used in the days of the horse-car. Of recent years, the most progressive roads have given much attention to the construction of equipment, in order to insure the safety of their passengers, and, as a result, braking appliances have been greatly improved.

The hand-brake was the first form of brake used on electric cars, and is still quite largely used. It is found to-day on most cars fitted with air-brakes, to be used in case of necessity. The early forms of the hand-brake consisted of a brake-staff located at either end of the car, having a chain connected to the lower end of the staff. As the handle is turned, the chain is wound up on the staff, and the resulting motion actuates the rods and levers which bring the brake-shoes in contact with the wheels. An improved form of brake-staff is that shown in Fig. 72. Here the winding drum takes the form of a spiral cam. In operation, the slack in the chain is quickly taken up and a very great braking pressure can be obtained.

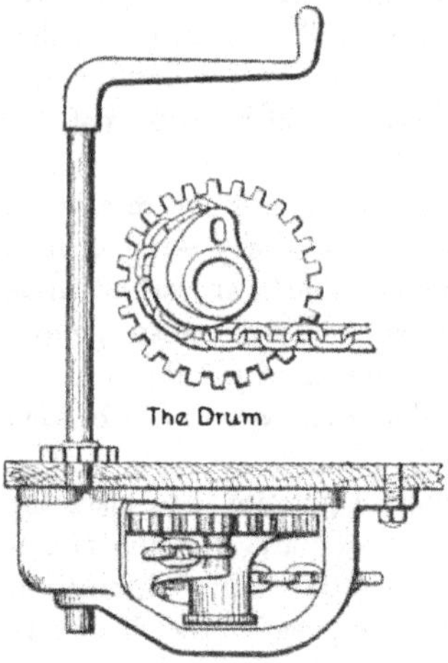

Fig. 72. Hand-Brake for Electric Cars.

The first form of air-brake installed on electric cars was the *Straight Air-Brake* system. It is largely used to-day, as is also the *Automatic Air-Brake* system. The Straight Air-Brake system is usually found on trains of not more than two cars in length. Since electric roads do not at this time interchange cars to any great extent, there is no very great necessity for interchangeable air-brake apparatus. As a result, there are a number of different types of air-brake apparatus found in use on electric cars. All operate more or less upon the same general principles.

As the space allotted to this subject is limited, only one system will be described, namely—the Westinghouse system. This system is chosen, since it represents in a general way other systems in use on many roads.

Westinghouse Straight Air-Brake. The action of the Westinghouse Straight Air-Brake system for electric cars is the same as that already described for steam roads (see page 4).

The system is composed of the following principal parts:

1. An *air-compressor*, operated by an electric motor, to provide compressed air.
2. A *governor* which automatically controls the action of the compressor, thereby maintaining the supply of compressed air at the proper pressure.
3. A *system of wiring*, with the proper switches, fuse-boxes, etc., which connect the trolley current to the governor and compressor.
4. A *large reservoir* in which compressed air is stored.
5. A *brake-cylinder* and *piston*, the piston-rod of which is connected to the brake-rods in such a manner that when compressed air is admitted to the cylinder, and the piston moves outward, the brake-shoes are pressed against the tread of the wheels.
6. An *operating valve* placed at either end of the car, by means of which compressed air can be admitted from the reservoir into the brake-cylinder and exhausted from the brake-cylinder to the atmosphere.
7. A *system of piping* connecting the above-mentioned parts, and, when trailers are used, including flexible hose and couplings and cut-out cocks.
8. A *safety-valve* connected to the reservoir to prevent too great an accumulation of air should the governor fail to operate.
9. A *chime whistle* connected to the air-supply, to be used as a warning of approach.

The general arrangement, names, and relative location of all parts, are shown diagrammatically in Fig. 73.

Operating the Straight Air-Brake. The operating valve has notches placed upon it which mark the position of the handle for the various positions of the valve. This fact enables one to operate the brake with certainty the first time, but smooth and accurate stops can be made only after a little practice. Beginning from the right and going to the left, the different positions of the valve handle are as follows: *Emergency position, service position, lap position,* and

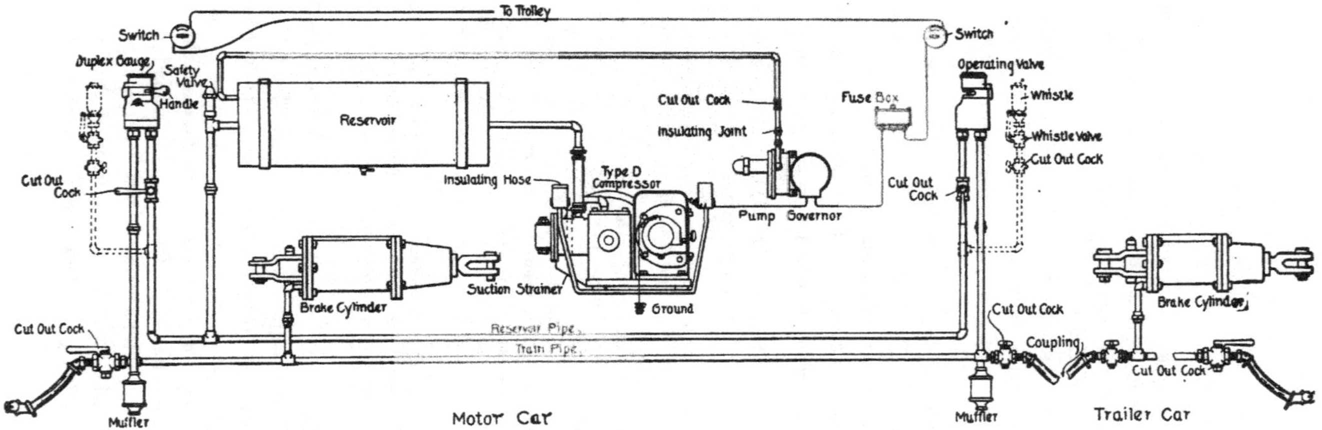

Fig. 73. Diagram of Westinghouse Straight Air-Brake System.

release position (see Fig. 83). When the handle is in the lap position, as indicated by the deep notch, the main ports in the valve are closed, and compressed air cannot enter the brake-cylinder from the reservoir, and any compressed air which may be in the brake-cylinder cannot exhaust into the atmosphere. If the handle is now moved from this position to the extreme left, it will then occupy the release position. In this position, any air which may have been in the brake-cylinder will be exhausted into the atmosphere, and the brake will be released. This is the position the handle should occupy while running on a level track. If the handle is moved from the release position to the service position, air will flow very slowly from the reservoir into the brake-cylinder, and service application results. If, however, the handle is moved from the release position to the extreme right (the emergency position), a large amount of air rushes from the reservoir into the brake-cylinder, and an emergency application is obtained. If the car is coasting down a grade, and the handle is moved to the service position for an instant and immediately returned to lap position, a small amount of air is admitted to the brake-cylinder and retained, thus holding the brakes applied. With a little experience, the proper amount of air can be admitted to the brake-cylinder in order that a constant speed may be maintained. If too much air is admitted into the brake-cylinder, a small portion can be exhausted by throwing the handle to release position for an instant, then back to lap position.

The quickest stop possible is made by throwing the handle at once to the emergency position, giving to the wheels the greatest possible braking pressure. The higher the speed, the greater the pressure that can be applied without danger of sliding the wheels. Thus it is seen that the quickest stop can be made by applying at once full braking pressure (depending on the speed), and gradually releasing as the speed decreases. This method insures a smooth stop, as the rapid reduction of speed at the end of the stop, which throws passengers forward, is avoided. In making a service stop, about twenty-five or thirty pounds of air-pressure should be quickly admitted to the brake-cylinder, and gradually reduced as the speed decreases, retaining about ten pounds in the cylinder until the car stops. A little experience is necessary in order to know just what pressures to use to be able to stop in a given distance. A succession of applications and release in stopping a car imparts a very disagreeable motion to the car, and is very wasteful of compressed air. In making emergency applications, the handle is thrown to the emergency position, and brake-cylinder pressure of, say, 60 pounds is obtained almost instantly. Sand should then be applied, and the handle be brought at once to the

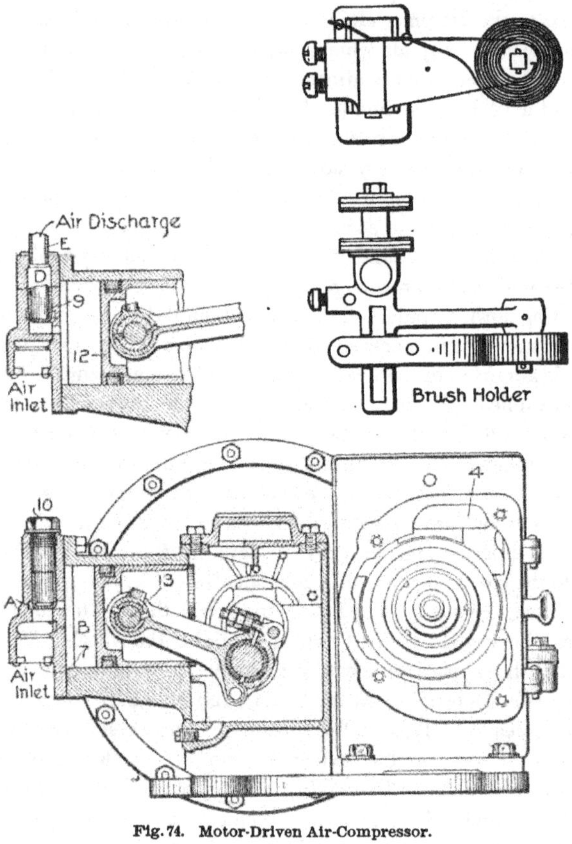

Fig. 74. Motor-Driven Air-Compressor.

lap position. The brake-cylinder pressure should then be released little by little as the speed drops.

When the signal is received to go ahead, the handle should be placed in release position before turning on the power. When descending a grade, the inexperienced man usually makes the mistake of applying the brake too hard at the start. It should be borne in mind that the car will not at once take the speed desired, and that some time is required for conditions to become constant. An easy application should first be made, and the handle held on *lap* until the car has sufficient time to feel the effect of the brake. If the speed of the car is still too high, let in a little more air, and repeat the operation as often as is necessary until off the grade.

The following instructions are given by the Westinghouse Company to motormen:

"When leaving the car, always set up the hand-brake, as some one might tamper with the cut-out cocks. Before starting from the car-barn, be sure all cocks are properly set, and that there is a good supply of air in the reservoir. Insert the handle in its socket in the operating valve, and throw it around to emergency, then back to release, to see that it works freely. Try the air-brake both in *service* and in *emergency*, to make sure that it has not been left improperly connected, etc. After this trial, and as long as proper pressure is maintained, the brake may be relied upon to perform its duty."

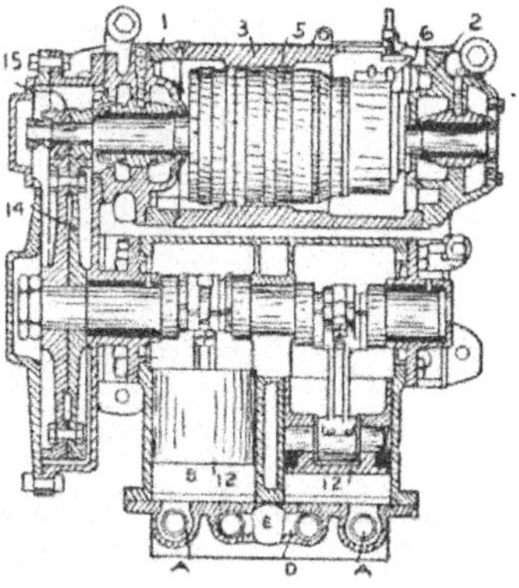

Air-Compressor. The air-compressor may be either axle-driven or motor-driven. Since there are some objections raised against using the axle-driven compressor, and since the motor-driven compressor is more commonly used, it is deemed advisable to confine ourselves to the motor-driven compressor. Reference will be made to Figs. 74, 75, and 76. All metal parts, such as pistons, rods, frames, etc., will be referred to as 1, 2, 3, etc., while all cavities and chambers will be called *A*, *B*, *C*, etc.

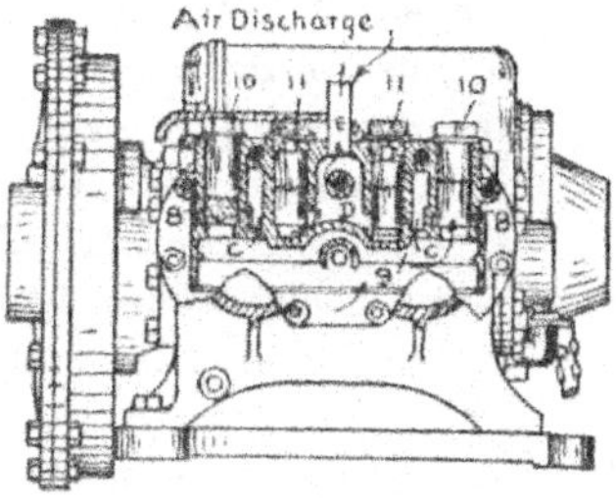

Fig. 75. Motor-Driven Air-Compressor.

The motor is of the series type, having an opening at the commutator end which permits of ready access to the commutator. This opening is provided with a tight-fitting door which excludes all dirt, dust, and moisture. In the ends of the frame are fitted heads (1) and (2), which provide bearings for the ends of the armature. Each bearing is provided with two oil-rings which secure proper lubrication of the shaft. Oil-holes are provided for filling the oil-wells, and the location is such that there is no danger of flooding the interior of the

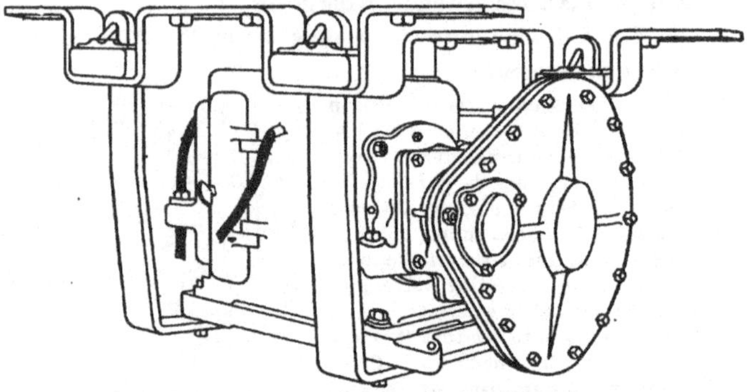

Fig. 76. Air-Compressor Suspended in Cradle under Car.

motor with oil. A passageway at the pinion end conducts any excess of gear-lubricating oil to the bottom of the gear case, thus assisting in preventing any flooding of the motor. Two of the poles of the motor are a part of the frame (3), and two are made up of soft laminated iron (4) bolted to the frame. The armature (5) is made up of soft-steel punchings which have accurately spaced slots in which are imbedded coils of uniform size. The brush-holders (6) are made of brass and are bolted to a cast-iron yoke with proper insulation. The brushes are carbon, and are held against the armature by coiled springs.

The action of the compressor in compressing air is as follows: Air is drawn through the suction screen (7), lifts the check-valves (8), and passes through the ports *A* into the cylinders *B*. On the return stroke, the compressed air is forced out through the ports *C*, lifting the discharge valves (9), then passing into the chamber *D*, and finally into the discharge pipe *E*. The suction and discharge valves are made of steel, and are accessible by removing the caps (10) and (11), respectively. These valves do not have any coiled springs to seat them, but close by gravity.

The pistons (12) are accurately fitted with rings, and are made long so as to reduce the amount of wear. When repairing the pump, the rings should always be kept with the piston to which they belong. The wrist-pin (13) is made of steel, and works on a bronze bushing in the connecting rod. The crank end of the connecting rod is lined with babbitt which works on the crank, and has suitable means for adjustment. The center line of the cylinders is placed above that of the crank-shaft, in order that the angularity of the connecting rod may be reduced during the compression stroke. This reduces the vertical component of the thrust on the pistons, and thereby reduces the amount of wear on the cylinders. It should be remembered, however, that the pump should always run with the compression part of the stroke on the upper half of the revolution. The crank-shaft is made of forged steel, and has two bronze bearings, one at either end, and a babbitt bearing in the middle. The crank-shaft bearings, wrist-pins, and crank-pins are lubricated by the splash system, from a bath of oil in the crank-case. The gear wheels (14) and (15) are of the *herringbone* type, and are lubricated from a bath of oil in the dust-proof gear-case.

An air-compressor heats very rapidly when in operation, if no means are provided to conduct the heat away. For this reason, compressors which are designed for continuous service are always water-jacketed. Since compressors for electric-car service are used intermittently, they have time to cool, and a water-jacket is unnecessary. Experience has shown that such a compressor as just described, when compressing air at 100 pounds per square inch, should not run longer than 15 minutes at a time, and should then be permitted to cool at least fifteen minutes. This compressor is suspended under the car, in the position shown in Fig. 76, when in service. The method of suspension permits of its being readily removed for repairing.

Pump-Governor. The location of the governor is shown in Fig. 73. Its purpose is to start and stop the compressor in order to maintain a predetermined pressure, by alternately making and breaking the circuit leading to the motor. A front view of the governor is shown in Fig. 77, and sections are shown in Figs. 78 and 79. The chamber *A* is in communication with the reservoir. The other side of the diaphragm (1) which forms one wall of the chamber *A* is open to the atmosphere. The diaphragm (1), therefore, is subjected to reservoir pressure on one side and atmospheric pressure and the regulating spring (2) on the other. The slide-valve (3) is connected to the diaphragm (1) in such a manner that any movement of the latter operates the former. When the maximum pressure is attained, the

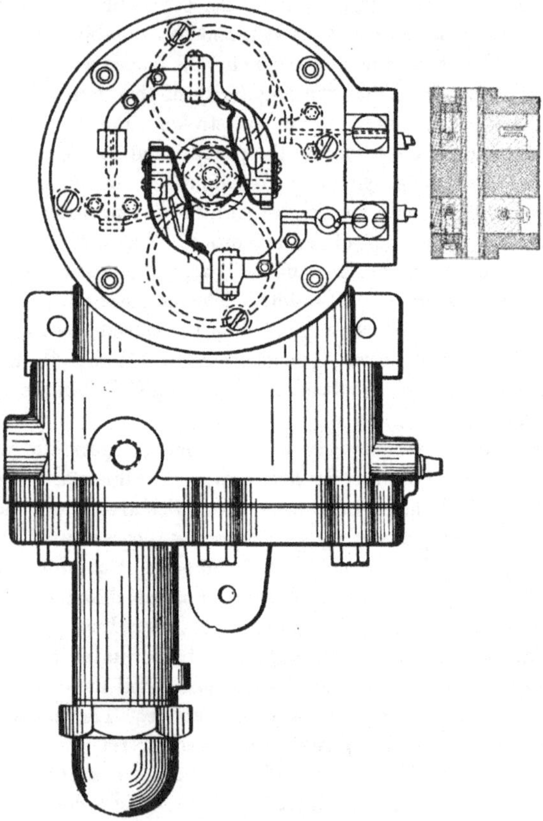

Fig. 77. Pump-Governor for Air-Compressor, Front View.

regulating spring (2) is so adjusted that the diaphragm (1) is pressed downward. This moves the slide-valve (3) and uncovers the port *B*, which is in communication with the chamber *C*. The air-pressure now in the chamber *C* forces the piston (4) upward, thereby opening the switch in the motor circuit, and the motor stops. When the air-pressure in the reservoir drops slightly, and consequently the pressure above the diaphragm (1) is reduced, the regulating spring (2) forces the diaphragm upward, which also moves the slide-valve (3) and connects the port *B* with the exhaust port *D*. Air from the chamber *C*

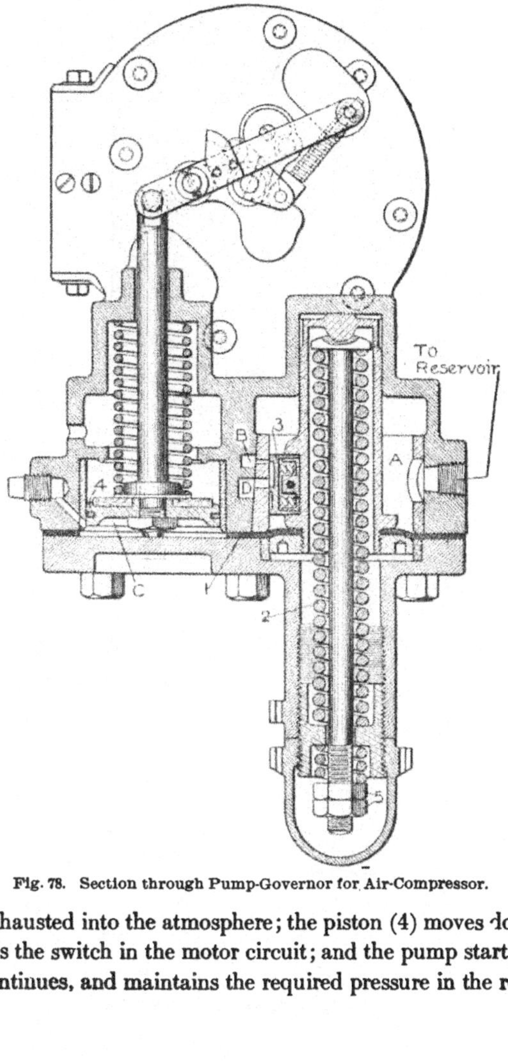

Fig. 78. Section through Pump-Governor for Air-Compressor.

is now exhausted into the atmosphere; the piston (4) moves downward and closes the switch in the motor circuit; and the pump starts. This action continues, and maintains the required pressure in the reservoir.

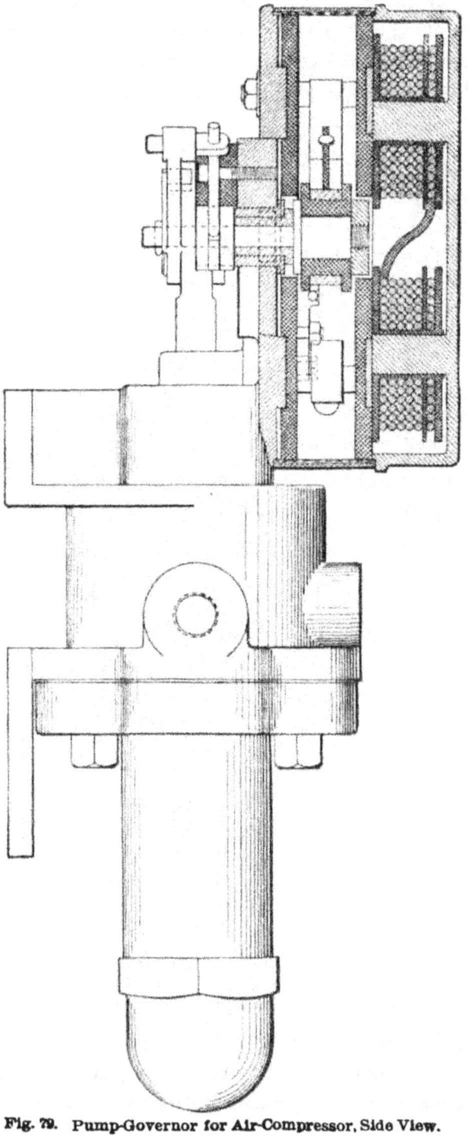

Fig. 79. Pump-Governor for Air-Compressor, Side View.

The mechanism on the upper part of the governor acts so as to cause the switch to open and close very rapidly, and thus avoids undue arcing. The pressure at which the governor cuts out the motor is controlled by adjusting the regulating spring (2) by means of the nuts (5). The governor may be located either under the car or in one end of the car.

The electric apparatus above described is for direct current. Alternating-current motors and governors are being used to some extent, but have not yet come into very general use.

Reservoir. The reservoir should have a sufficient capacity to supply air for three or four applications without reducing the pressure more than 15 pounds. It is conveniently located under the car, and its dimensions depend upon the size of the brake-cylinder used. It serves to collect moisture and oil, and prevents them from being carried further into the system. It should be drained frequently, as its capacity for stored air will be reduced proportionally to the volume of water it contains.

Brake-Cylinder. The brake-cylinder shown in Fig. 80 is of the

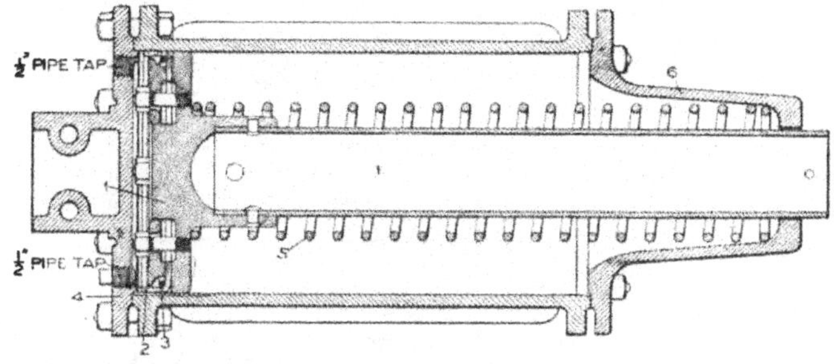

Fig. 80. Brake-Cylinder of Hollow-Rod Type.

hollow-rod type. The piston is connected to the brake-rigging in such a way that it moves only when the power-brake is used. When the hand-brake is used, no movement of the piston occurs. The piston rod (1) is made hollow to receive the push rod. A leather packing-ring (2) is provided which prevents air from leaking around the piston. The leather packing-ring is held against the walls of the cylinder by means of the round spring expander (3). The cylinder-head (4) may be either plain or as shown. That shown is constructed to receive an automatic slack-adjuster (see page 79), which is sometimes used with the automatic system.

When an application is made, air enters behind the piston and forces it outward, compressing the release spring (5). When the air is exhausted from the cylinder, the release spring (5) pushes the piston back to its normal position. Cylinder head (6) is constructed so as to provide a place for the coil spring when the piston is forced outward.

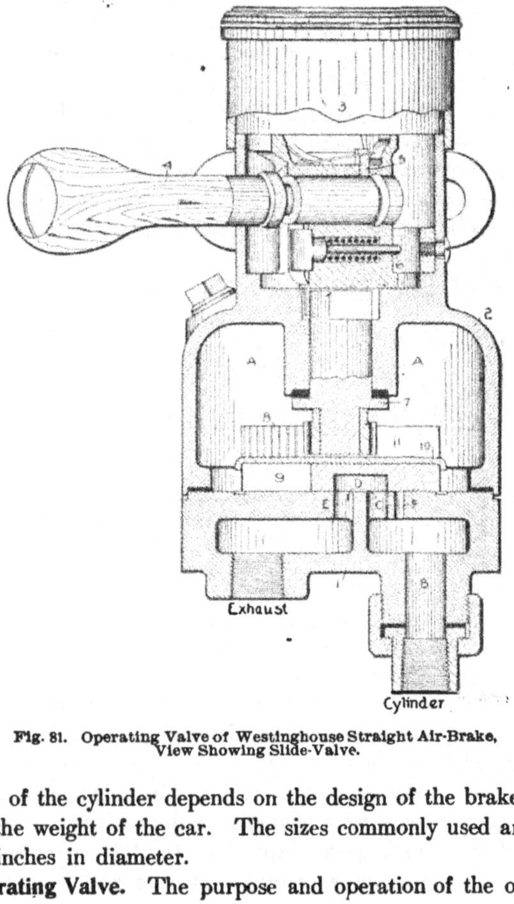

Fig. 81. Operating Valve of Westinghouse Straight Air-Brake, View Showing Slide-Valve.

The size of the cylinder depends on the design of the brake-rigging and on the weight of the car. The sizes commonly used are 8, 10, and 12 inches in diameter.

Operating Valve. The purpose and operation of the operating valve has already been described (see page 91). The four positions of the valve are: *Emergency, service, lap,* and *release.*

When in emergency position, full braking pressure is obtained almost instantaneously, and is used in avoiding collisions and making quick stops. In this position, direct communication is made between the reservoir and the brake-cylinder. The rail should always be sanded to avoid the possibility of slipping the wheel, which would result in making a poor stop and would probably cause a flat spot on the wheel.

In service position, air enters the brake-cylinder pipe through a small port in the operating valve, and applies the brake very slowly.

When in lap position, the ports in the operating valve are blocked, and air can flow neither to nor from the brake-cylinder. If the brake is applied, it will remain so until the valve is thrown to release position.

If the valve is placed in release position, the cylinder and exhaust ports are connected, and only atmospheric pressure will remain in the cylinder. If the brake has been applied, it will release when the valve is placed in this position.

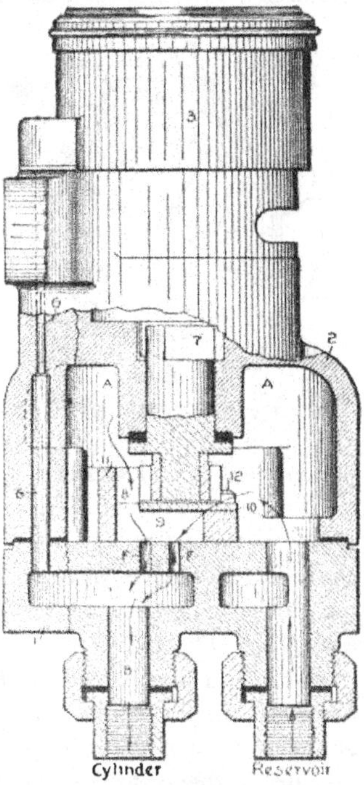

Fig. 82. Operating Valve of Westinghouse Straight Air-Brake, View Showing Reservoir Cavity.

. In describing the operating valve, reference will be made to Figs. 81, 82, and 83. The valve is cast in two parts—the base (1) and the head and body (2). On the top of the head is a double gauge (3); the red hand indicates the reservoir pressure; and the black hand, the brake-cylinder pressure. Just below the gauge is a socket into which fits the operating handle (4) which is removable. In swinging from release position to emergency position, the handle turns through about 130 degrees. The handle can be inserted and withdrawn only when the valve is in lap position. When the handle is withdrawn, the latch (5) is thrown into position by a small spring, and the valve is permanently locked until the handle is again inserted. Just below the handle socket is a second one which contains a bolt (6) actuated by a spring. As the handle is turned, the head of the bolt (6) passes over notches which serve to indicate when the valve is in the proper position. Connected to the lower side of the socket is the stem (7) having a

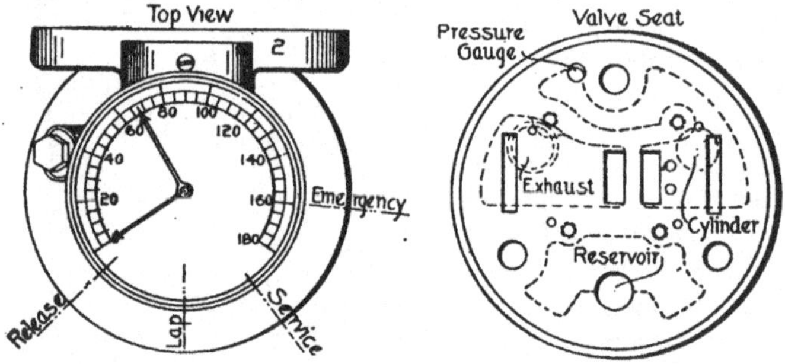

Fig. 83. Top View and Valve-Seat of Westinghouse Straight Air-Brake.

pinion fitted to its lower end, which actuates the rack (8). The rack (8) is connected to and operates the slide-valve (9). The spring plate (10) does not act as a stop for the slide-valve (9), but is used only to assist in getting the valve in the proper position when assembling the parts. The slide-valve (9) moves between suitable guides (11) and (12). The chamber *A* is always in communication with the reservoir, and a port leads to the gauge above, which indicates the pressure. In the figure, the valve is shown in release position; air passes from the cylinder through the pipe *B*, the port *C*, the cavity *D*, the port *E*, thence to the exhaust pipe. When the valve is in emergency position, the right-hand edge of the slide-valve (9) registers with the left-hand edge of the port *C*. Air then passes from the chamber *A*, through the ports *C* and *F*, through the pipe *B*, to the brake-cylinder. In this position, the port *E* is blocked. In lap position, the right-hand portion of the slide-valve (9) covers the ports *C* and *F*, and the port *E* is blocked. The port *G* connects the brake-cylinder pipe with the gauge above, which indicates the cylinder pressure.

Another form of this operating valve is sometimes used which has no gauge at the top to indicate the cylinder and reservoir pressures. The operation of the valve is the same as in the case of the one just described.

The valve just described is sometimes used in a modified form as shown in Fig. 84. Here the operating handle and valve parts are

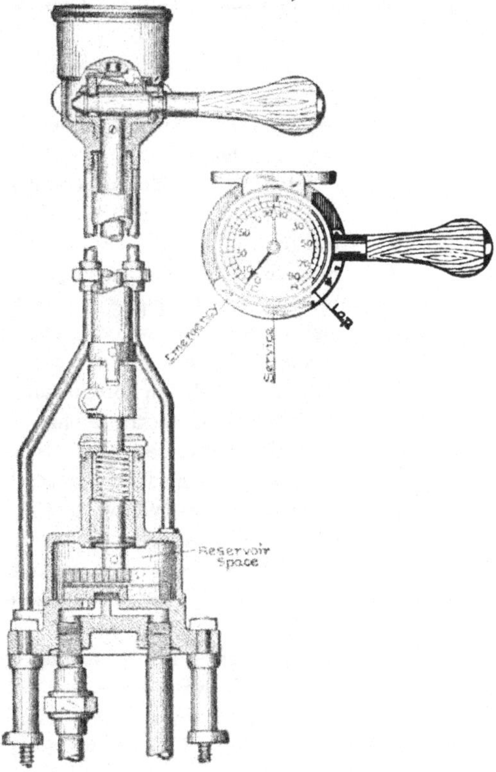

Fig. 84. Type of Operating Valve with Handle and Valve Parts Separate. Westinghouse Straight Air-Brake System.

separate, and the valve parts are bolted to the floor of the car. In operating this brake, the handle must be thrown in a way the reverse of that just described, but otherwise the operation of the valve is the same as previously given.

Piping. Referring to Fig. 73, the sizes of the various pipes are as follows:

The train-pipe connecting the brake-cylinder with the operating valve should be a standard $\frac{1}{2}$-inch pipe. If more than one trailer is used, a $\frac{3}{4}$-inch pipe should be used.

The reservoir pipe, connecting the reservoir with the operating valve is a $\frac{1}{2}$-inch pipe. A $\frac{3}{4}$-inch pipe is better if it can be used conveniently.

The pump-governor and whistle connections are made with $\frac{3}{8}$-inch pipes. Wherever possible, long bends in pipes should be used, rather than a standard elbow fitting.

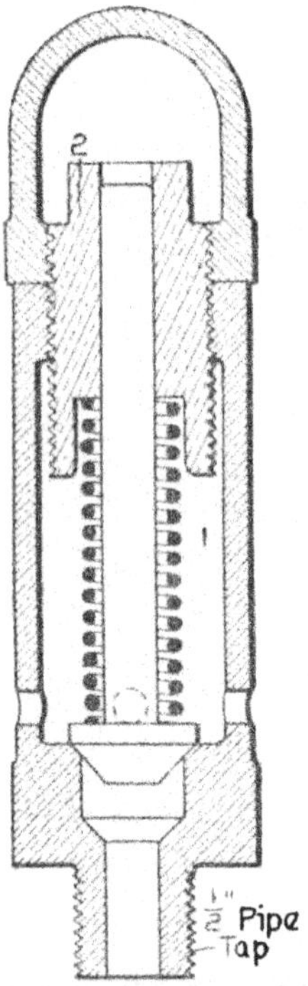

Fig. 85. Safety-Valve of Westinghouse Straight Air-Brake.

Safety-Valve. The safety-valve should be connected to the reservoir line leading to the controlling valve, at a point near the reservoir. Its operation may be understood by reference to Fig. 85. It can be set for any pressure by adjusting the regulating spring (1) by means of the nut (2).

In an axle-driven compressor equipment, a slight change in the piping is necessary from that above described. Since the compressor is mounted on the truck, and has some movement relative to the car frame which carries the reservoir, flexible hose connections are necessary, to make connections to the reservoir and also to the compressor regulator. A small reservoir is also used which receives air from the compressor. This small reservoir is connected to the main reservoir by a pipe containing a regulating valve. The air attains a pressure of about 35 pounds in the small reservoir before any air passes into the main reservoir. This 35 pounds' pressure in the small reservoir is attained while the car runs about 100 yards, and is available for applying the brakes. This always insures air for operating the brakes if the car previously runs a short distance. With this ex-

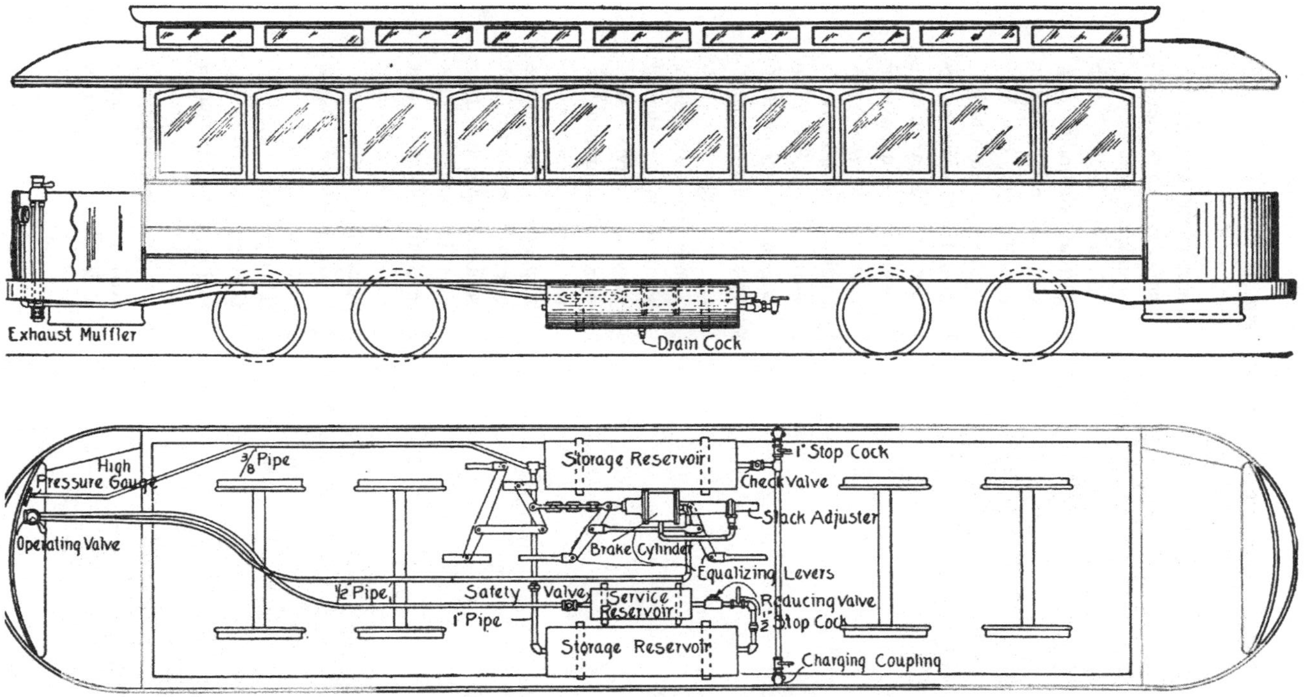

Fig. 86. General Scheme of Storage Air-Brake Equipment on a Car.

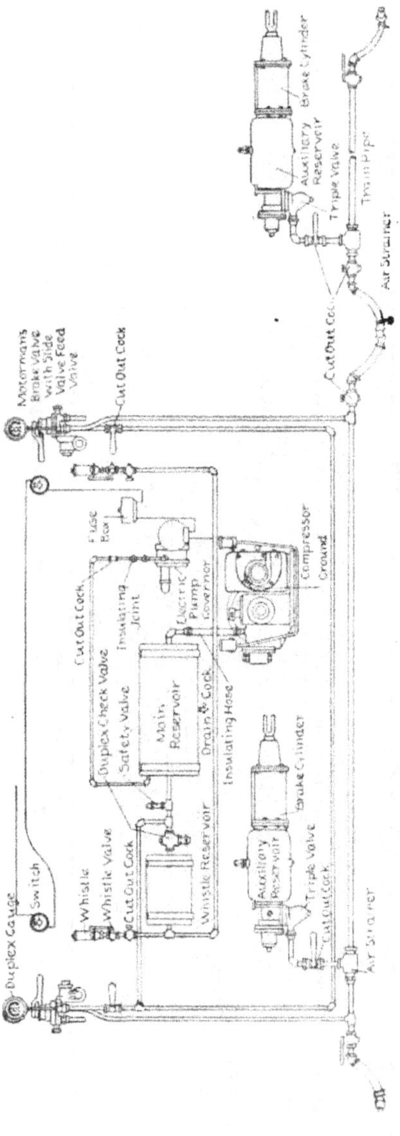

Fig. 87. Westinghouse Automatic Friction-Brake. Motor-car equipment shown at left; trailer-car equipment at right.

ception, the piping is the same, and no further description is necessary.

If a car is fitted with a storage air-brake equipment, no compressor is installed in the car. The compressed air which is used for braking is carried on the car in large reservoirs. The general scheme of a storage air-brake equipment is shown in Fig. 86. Two large reservoirs connected by a one-inch pipe carry air at high pressure. These reservoirs deliver air through a reducing valve to a service reservoir. The pressure in the service reservoir corresponds to that in the reservoir previously described. Other than these parts just mentioned, the straight air-brake and the storage air-brake systems are the same.

Westinghouse Automatic Friction-Brake. The general scheme of this equipment is shown in Fig. 87, which gives the names of the principal parts and their relative location. The principle of its operation is very different from that of the straight air-brake system. In the straight air-brake system, the brake-pipe is subjected to pressure only when an application is made. With the automatic system, air at 70 pounds' pressure per square inch is carried in the brake-pipe. The brake is applied by exhausting air from the brake-pipe, thus reducing its pressure; and it is released by restoring this pressure. It follows that any accident or operation which results in reducing the brake-pipe pressure will apply the brakes on all cars. This is not true, however, in case of the straight air-brake system. In the straight air-brake system, if any accident occurs to break or open the brake-pipe, the brake at once becomes inoperative. With the exception of compressed air being supplied by a motor-driven compressor, a governor controlling the operation of this compressor, and a change in the form of the brake-valve, the system is almost identical with the Westinghouse system already described for steam-operated roads. The descriptions of the operation of the automatic brake already given apply equally well to the automatic system for electric cars. The system is especially recommended for use on trains of more than two cars, where frequent stops are not required.

The standard automatic air-brake system as used on steam roads to-day cannot be successfully operated on electric trains for street service composed of one car, for the following reasons:

First. Applications of the brake are likely to follow in such rapid succession that sufficient time would not be given to properly recharge the auxiliary or braking reservoir on each car.

Second. A graduated release or gradual decreasing brake-cylinder pressure is absolutely necessary in electric-car work, in order to obtain a smooth stop. With the standard automatic equipment, release of the brake-cylinder pressure is complete, when once started.

Third. A prompt response of the brakes when re-applied after a release, is very essential. This is not always possible in the standard automatic equipment, since the auxiliary reservoir is very slow in charging.

To overcome these difficulties, there has been devised an automatic system for electric-car work, having quick-service, graduated-release, and quick-recharging features. This system is very important for a certain class of service, but will not be described.

Train Air-Signal. As the size of electric cars and the length of trains increase, a signal system becomes more and more a necessity. That used to-day on steam roads has been fully described in preceding pages. Since the air-signal system used on electric cars is the same as that used on steam roads, it is unnecessary to repeat the description.

Stopping a Car. The brake equipment of all electric cars is calculated with reference to the unloaded weight of the car, that is—the parts are so designed that there will be no danger of slipping the

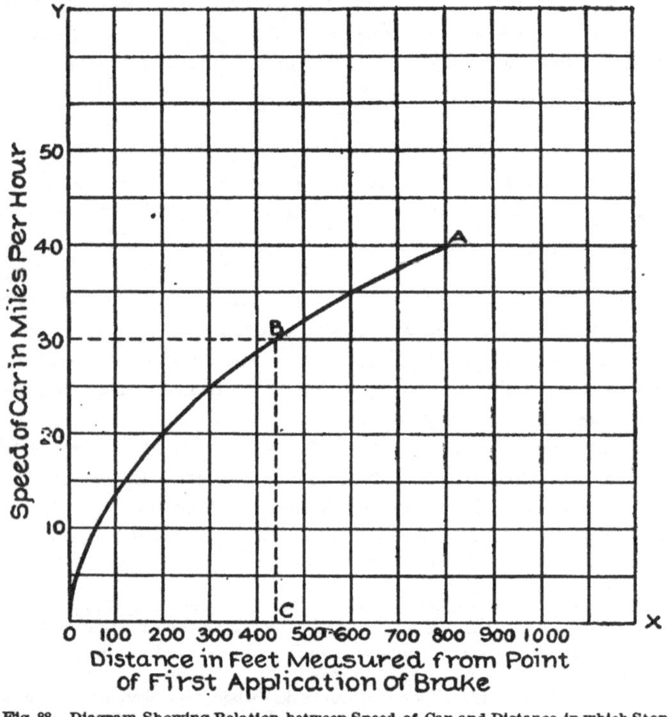

Fig. 88. Diagram Showing Relation between Speed of Car and Distance in which Stop can be Made after Application of Brake.

wheels when the car is unloaded. In stopping a car, the forces which act to retard its motion are:

The resistance of the atmosphere;

The frictional resistance of the journals and track; and

The resistance of the brake-shoes on the wheels.

When the brake is applied, the car pitches forward on the front truck, and the weight on the rear truck is thereby decreased. If proper allowances have not been made in proportioning the brake-levers, the rear wheels will probably slip on the track. If the wheels should slip, the distance required in which to bring the car to rest would probably be greater than that required had the wheels not slipped. In bringing a car to rest, the energy of translation of the entire car and the energy of rotation of all the wheels and motors must be absorbed by friction. To do this efficiently and safely in the shortest possible time, is the purpose of the modern brake systems.

The average person who rides on street and interurban cars knows nothing as to the distance in which these cars can be stopped. "In what distance can a modern double-truck electric car be stopped?" is a question which is frequently asked. In answer to this question, Fig. 88 has been prepared. A great many experiments have been made in stopping cars, with varying results. The chief factors which affect the results of such tests are the condition of the rail and the character of the material composing the brake-shoes. Fig. 88 shows graphically the relation between the distance required to stop a car and the speed (in miles per hour) at the instant the brake was applied. It represents the average result of a large number of experiments with a double-truck car fitted with brake equipment as described in the preceding pages. With perfect conditions, the curve *A B O* would fall above that shown, while with very poor conditions, it would fall lower. The value of the diagram is made apparent by the following application:

Example. Find the distance in which a double-truck electric car may be stopped if power is shut off and the brake applied while running at a speed of 30 miles per hour.

Solution. Starting on the vertical line *O Y* at 30 miles per hour, follow the horizontal line to the right until the curve *A B O* is reached at the point *B*. From the point *B*, follow the vertical line downward until the horizontal line *O X* is reached at the point *C*. This point *C* indicates the distance in feet in which the car may be stopped, which in this instance is 440 feet. In the same way, the stopping distances may be determined for cars running at any speeds.

INDEX

INDEX

The School Behind the Book

THIS practical handbook is one of the representatives of the American School of Correspondence. It is the only kind of representative by which the School reaches the general public and extends its educational work.

The American School of Correspondence is chartered, under the same laws as a State University, as an educational institution. Its instruction books, written especially to suit the needs of men seeking self improvement through correspondence work, are reserved for its students and for class use in educational institutions; many of these texts are used in the class room work of the best resident schools in the country.

However, in order that the large number of ambitious men, for whom class work and correspondence study are neither practical nor advisable, may not be deprived of this valuable material, it is published by the School both in sets covering the several branches that it teaches, and in a series of single Home Study volumes treating of specialized lines of practical knowledge. This book is a sample of the make-up of the Home Study volumes and the titles and authors are shown on the following page. By this method the School broadens its field of activity; and from these sales it derives an income to use in general educational work.

The School's publications are clear and practical, and will be found ideal for reference and home reading. For those, however, who desire more systematic study of the subjects in which they are particularly interested, the School advises a thorough course by correspondence as the quickest and surest means of obtaining the practical knowledge desired.

The School offers correspondence instruction in all branches of architecture, civil engineering, college preparatory work, accounting and business administration, drawing and design, electrical engineering, fire prevention and insurance, American law, mechanical, sanitary, and steam engineering, and textile manufacturing. It adapts its courses to the needs of the individual, by starting him where his previous education stopped, and giving him only such work as is necessary to fit him for the work he wants to do.

On request the School will mail to any address a Bulletin containing full information regarding its courses and methods. It employs no representative other than its own publications.

Partial List of Titles and Authors—Continued

		PRICE
Electric Wiring and Lighting	Knox-Shaad	$1.00
Estimating	Edward Nichols	1.00
Factory Accounts	Hathaway-Griffith	1.50
Forging	John Lord Bacon	1.00
Foundry Work	Wm. C. Stimpson	1.00
Freehand and Perspective Drawing	Everett-Lawrence	1.00
The Gasoline Automobile	Lougheed-Hall	2.00
Gas Engines and Producers	Marks-Wyer	1.00
Heating and Ventilation	Charles L. Hubbard	1.50
Highway Construction	Phillips-Byrne	1.00
Hydraulic Engineering	Turneaure-Black	3.00
Insurance and Real Estate Accounts	Charles A. Sweetland	1.50
Knitting	M. A. Metcalf	3.00
Machine Design	Charles L. Griffin	1.50
Machine-Shop Work	Frederick W. Turner	1.50
Masonry and Reinforced Concrete	Webb-Gibson	3.00
Masonry Construction	Phillips-Byrne	1.00
Mechanical Drawing	Ervin Kenison	1.00
Modern American Homes	H. V. von Holst	3.00
Motion Pictures	David S. Hulfish	4.00
The Orders	Bourne-von Holst-Brown	3.00
Pattern Making	James Ritchey	1.00
Plumbing	Gray-Ball	1.50
Power Stations and Transmission	Geo. C. Shaad	1.00
Practical Aeronautics	Chas. B. Hayward	3.50
Practical Bookkeeping	James B. Griffith	1.50
Practical Lessons in Electricity	Millikan-Knox-Crocker	1.50
Reinforced Concrete	Webb-Gibson	1.00
Railroad Engineering	Walter Loring Webb	3.00
Refrigeration	M. W. Arrowwood	1.00
Sewers and Drains	A. Marston	1.00
Sheet Metal Work	William Neubecker	3.00
Stair-Building and Steel Square	Hodgson-Williams	1.00
Steam Boilers	Newell-Dow	1.00
Steam Engines	L. V. Ludy	1.00
Steam Turbines	Walter S. Leland	1.00
Steel Construction	E. A. Tucker	1.50
Strength of Materials	Edward Rose Maurer	1.00
Surveying	Alfred E. Phillips	1.50
Telephony	Miller-McMeen	4.00
Textile Chemistry and Dyeing	Louis A. Olney	3.00
Textile Design	Fenwick Umpleby	3.00
Tool Making	Edward R. Markham	1.50
Valve Gears and Indicators	L. V. Ludy	1.00
Water Supply	Frederick E. Turneaure	1.00
Weaving	H. William Nelson	3.00
Wireless Telegraphy and Telephony	Ashley-Hayward	1.00
Woolen and Worsted Finishing	John F. Timmerman	3.00
Woolen and Worsted Spinning	Miles Collins	3.00

Spotted something that isn't quite right?

Let us know and we'll fix it for future reprints.
And as a thank you for your diligence, we'll
send you a free Sothis book of your choice.

publisher@sothis.press

This page intentionally left blank

Except of course for that line above. And this one.
And whilst we have your attention, perhaps you'd
like to follow us on Instagram @sothis.press. Go on.

Made in the USA
Las Vegas, NV
02 May 2026

46291341R20080